SUSTAINABLE VENICE: SUGGESTIONS FOR THE FUTURE

FONDAZIONE ENI ENRICO MATTEI (FEEM) SERIES ON ECONOMICS, ENERGY AND ENVIRONMENT

This series serves as an outlet for the main results of FEEM's research programmes in the areas of economics, energy and environment.

---

1. C. Carraro and D. Siniscalco (eds.), *The European Carbon Tax: An Economic Assessment.* 1993 ISBN 0-7923-2520-6
2. C. Carraro (ed.), *Trade, Innovation, Environment.* 1994 ISBN 0-7923-3033-1
3. C. Dosi and T. Tomasi (eds.), *Nonpoint Source Pollution Regulation: Issues and Analysis.* 1994 ISBN 0-7923-3121-4
4. C. Carraro, Y. Katsoulacos and A. Xepapadeas (eds.), *Environmental Policy and Market Structure.* 1996 ISBN 0-7923-3656-9
5. C. Carraro and A. Haurie (eds.), *Operations Research and Environmental Management.* 1996 ISBN 0-7923-3767-7
6. I. Musu and D. Siniscalco (eds.), *National Accounts and the Environment.* 1996 ISBN 0-7923-3741-7
7. C. Carraro and D. Siniscalco (eds.), *Environmental Fiscal Reform and Unemployment.* 1996 ISBN 0-7923-3750-6
8. A. Beltratti, *Models of Economic Growth with Environmental Assets.* 1996 ISBN 0-7923-4032-9
9. G. Chichilnisky, G. Heal and A. Vercelli (eds.), *Sustainability: Dynamics and Uncertainty.* 1998 ISBN 0-7923-4698-X
10. R. Roson and K.A. Small (eds.), *Environment and Transport in Economic Modelling.* 1998 ISBN 0-7923-4913-X
11. A. Markandya and M. Pavan (eds.), *Green Accounting in Europe - Four Case Studies.* 1998 ISBN 0-7923-5470-2
12. A. Lanza (ed.), *Resources Accounting in China.* 1999 ISBN 0-7923-5485-0
13. C. Carraro (ed.), *International Environmental Agreements on Climate Change.* 1999 ISBN 0-7923-5515-6
14. C. Carraro and F. Lévêque (eds.), *Voluntary Approaches in Environmental Policy.* 1999 ISBN 0-7923-5516-4
15. C. Carraro (ed.), *Efficiency and Equity of Climate Change Policy.* 2000 ISBN 0-7923-6262-4
16. I. Musu (ed.): *Sustainable Venice: Suggestions for the Future.* 2000 ISBN 0-7923-6458-9

KLUWER ACADEMIC PUBLISHERS – DORDRECHT / BOSTON / LONDON

Sustainable Venice: Suggestions for the Future

Edited by

IGNAZIO MUSU

University of Venice, Italy and Fondazione Eni Enrico Mattei, Milan, Italy

KLUWER ACADEMIC PUBLISHERS
DORDRECHT / BOSTON / LONDON

Library of Congress Cataloging-in-Publication Data is available.

ISBN 0-7923-6458-9

Published by Kluwer Academic Publishers BV,
PO Box 17, 3300 AA Dordrecht, The Netherlands.

Sold and distributed in North, Central and South America
by Kluwer Academic Publishers,
101 Philip Drive, Norwell, MA 02018, USA

In all other countries, sold and distributed
by Kluwer Academic Publishers, Distribution Center,
PO Box 322, 3300 AH Dordrecht, The Netherlands

Printed on acid-free paper

Printed and bound in Great Britain by MPG Books Limited, Bodmin, Cornwall.

Table of Contents

CHAPTER 1

Venice and its Lagoon: A Problem of Local Sustainable Development

Ignazio Musu

1. The lagoon and the sustainable development of Venice

Man is a biological species with peculiar characteristics within the ecosystem of which he is a part; this is because he possesses tools of interaction with the environment which no other species can use. These tools allow him to alter the balance between the evolutionary developments of his own history and the development of the biological ecosystem. This capacity to alter, which is connected to the discrepancy between historical timescales and biological timescales, lies at the root of the problem of sustainable development. However, the capacity to control this alteration, by reducing the results of the discrepancy to a minimum, is also the key to solving the problem.

The relationship between human community and environment can be distinguished in different typologies and according to different historical events, each of which is itself the product of a previous development in the relationship between man and environment. The problem of sustainable development must be tackled taking these historical characteristics into consideration. Hence, the 'Venice case' is an emblematic example of the complex interaction between economy, society and environment.

A debate on the future of Venice has been underway for years (UNESCO, 1969; Dorigo, 1972; Costa, 1993). As a result of this debate, an awareness has developed of how central the problem concerning the relationship between development and the environment is for the future of the city. In recent years, the need for sustainable development emerged out of this awareness. In the Venetian case, the problem of sustainability can be specifically defined as the problem of the relationship between economic development and the lagoon ecosystem. Today, this relationship is in a problematic situation because of the specific historical evolution undergone by these two components.

I. Musu (ed.), Sustainable Venice, 1–26.

Venice is a city made up not only of its historical centre, but also of the mainland, the lagoon islands and the coast. The lagoon is the unitary environmental reference context for urban Venice and is, therefore, the term of comparison which must be favoured in the analysis of the environmental sustainability of local development. Although the historical, artistic and monumental environments of the historical centre take on particular importance, the peculiarity of the latter is still determined precisely by the fact that it is built on water and, therefore, on the lagoon. The problem of the sustainability of development is embodied in the possibility of creating and maintaining a model of economic activity and human life which is structured in harmony with, and not in opposition to, this complex lagoon environment. The environmental problem is becoming increasingly perceived as one which goes beyond the reduction in pollution; it also involves the maintenance and preservation of a whole set of natural values which find their unique expression in the lagoon ecosystem.

Today, economic development no longer ends in industrial development alone. The growth of tourism has taken the pole position in the economy of the historical centre and the coast, and also constrains the economic development of the mainland. The problem of Venice's position in post-industrial society emerges powerfully. On the other hand, technological progress has considerably increased the degree of industry's potential for compatibility with the environment.

One must not make the mistake of believing that without Venice the lagoon would be an ecosystem with greater quality and value than the present one, since without Venice, but with an intense anthropisation of the mainland, sediments would have caused the lagoon to disappear sooner or later. Not only has the existence of Venice saved the lagoon, but it has transformed the lagoon into a wetland which is unique in the world, precisely because it not only contains a natural and an architectural heritage but is also a living city. This fragile equilibrium between a city lived in by people, urban structures and environment is a specific objective of the sustainable development of Venice: it has been and will always be an unstable equilibrium which is both subject to dangers and open to opportunities.

The instability of the equilibrium is derived from the characteristics of the lagoon environment (UNESCO, 1995; Rinaldo, this volume). Indeed, the lagoon is an extraordinary environment which is, by definition, in a state of unstable equilibrium due to the combination of two groups of factors: those coming from the land and those from the sea. Their action allows it to exist and, at the same time, threatens its very existence. The rivers, carrying detritus and sediments, pose the threat of filling it in (to which the Venetian Republic reacted by diverting the rivers into the sea); but, if too much space is left to the excavating force of the sea's currents, the erosion processes threaten to turn the lagoon into a reach in the long-term.

To these two groups of factors, the action of the phenomena of subsidence and eustatism can be added; that is, the lowering of ground level and the rising

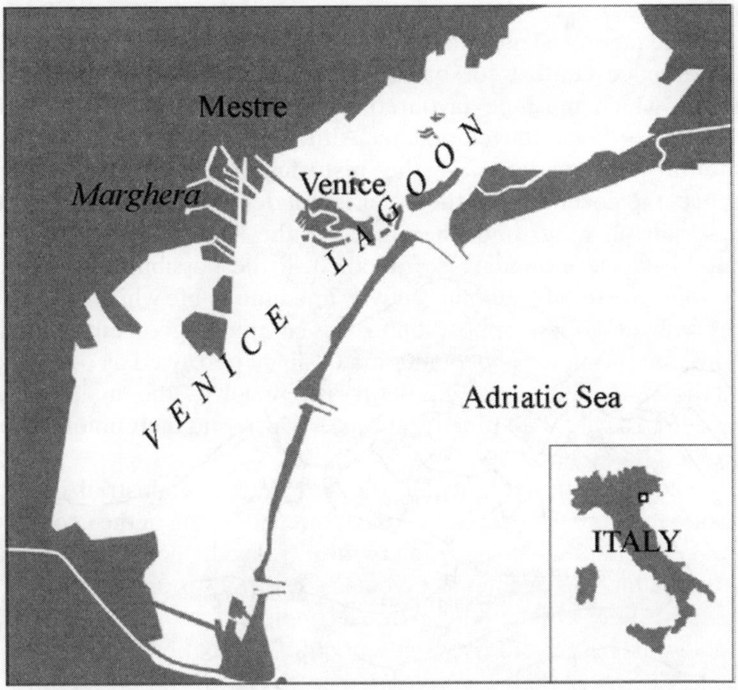

Figure 1. The Venice Lagoon

of sea-level. The increase in the average ratio of water level to ground level, which can be attributed mainly to subsidence (lowering of ground level) for the past, but also to eustatism (rise in sea-level), has caused an increase both in the average number of times which the historical city of Venice is struck by the problem of high water as well as in the average level of the high tide itself. The modifications in the hydrodynamic structure of the lagoon, which have taken place and which continue to occur as a result of human intervention in the lagoon itself, add to these basic factors and thereby contribute to exacerbating the effects of the phenomenon of high water (Rinaldo, this volume).

Venice has lived with this phenomenon throughout the centuries but the city has also been designed with the objective of reducing its impact on it to a minimum. Subsidence in the past and the threat of eustatism in the future has provoked a crisis in the relationship between the city and high water. The latter has and will become more frequent and the likelihood of exceptional events will increase. This not only exacerbates the damage to the physical structures of the city, thereby increasing the need for very expensive maintenance and architectural restoration work, but will also create increasing problems in maintaining a normal lifestyle in the city. A possible solution to this problem, involving some kind of flexible separation of the lagoon from the sea, would,

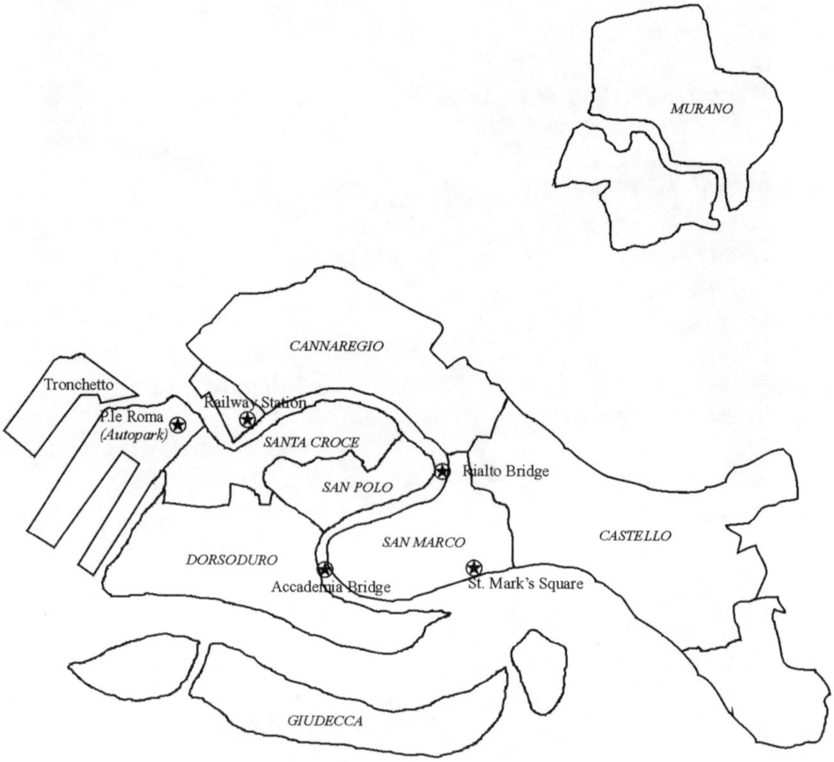

Figure 2. A simple map of Venice

however, also have some impact on economic activities and on the environment. The reduction of this impact to a minimum is one of the most important challenges for the sustainable development of Venice and its lagoon.

Throughout history, the criteria to be adopted for the maintenance of the lagoon were the subject of debates which obsessed and divided the ruling class of the Venetian Republic. This means that the statement 'the lagoon must be maintained', alone has no operational meaning. The very objective of maintaining the lagoon has been subject to different interpretations for the last 400 years. However, until the 1800s, differing opinions had not yet caused the breakdown in the historical objective of assigning a dual function to the lagoon; i.e. of satisfying a civil function (the protection of Venice and also its military defence) at the same time as an economic one (the development of the Republic's commercial power).

Venice's entry into the industrial era, beginning from the second half of last century, began to establish a distancing between the objective of maintaining the lagoon, a goal of a typically environmental and hydraulic nature, and the objectives of economic and social development. Today, the likelihood of possible conflict between objectives of maintaining the lagoon ecosystem and objectives of economic development is increasing. This is, above all, the result of the increase in erosion, in pollution and in the threat of an intensification of eustatism due to the effects of global warming on sea level.

On the other hand, the development of technology for immaterial production and the modern economy's new specialisation in the service sector make it possible to expand the range of routes for the city's economic specialisation, so as to not only render development sustainable from an economic perspective, but also to make it compatible with the objective of maintaining the lagoon as an ecosystem endowed with natural features which qualify it as such.

The modifications made in the morphology and, in particular, the erosive processes have had an impact on the natural lagoon environment (Torricelli *et al.*, this volume). The amount of the lagoon's surface area which is occupied by the valuable natural structures which correspond to the Venetian names *barene* (salt marshes), *ghebi* (little channels), and *velme* (shoals) has decreased from 20% at the beginning of the century to less than 8%. This reduction is linked to the above-mentioned effects of erosion, but also to past reclamation works, areas of which are presently undergoing natural restoration.

However, a certain recovery of the biodiversity in the Venice lagoon is taking place, especially in terms of the presence of birdlife (Torricelli *et al.*, this volume). The current positive trends are certainly the result of increased concern on the part of the public and of institutions for the protection of the features of the natural lagoon environment. Over the last few years various initiatives have been undertaken in order to create a comprehensive potection of the lagoon environment.

In spite of the fact that for 20 years Italy has been a signatory to the Ramsar Convention on the protection of wetlands of international importance, the Venice lagoon is still not wholly included in the list of such wetlands. The lagoon largely satisfies the two objective criteria which determine the international importance of a wetland, according to the Ramsar Convention. These criteria are to host at least 20,000 aquatic birds on a regular basis or to host at least 1% of the individuals of a world-wide population of any species of aquatic bird in any period of its annual biological cycle. The results of censuses on avifauna demonstrate that the first threshold is easily passed and that five species have passed the second (Torricelli *et al.*, this volume).

It must not be forgotten that the lagoon constitutes an example of an anthropised wetland. Therefore, its future as an area of natural interest to be protected has to be constructed by taking into consideration a plan for sustainable economic development for the human community residing there. In one way, this establishes a series of limits for this plan. From another complementary perspective, the lagoon, as an area of natural value, can be considered a resource within the plan for sustainable development. Environmental protection is a general objective for the whole Venetian lagoon, but some areas of it, within a comprehensive plan, have been made the object of special natural protection. In these areas, a kind of activity with economic value which makes use of as well as helps the protection of nature could well be developed.

The directions in which planning and initiatives should be oriented are those of sustainable natural tourism, of an economic activity of sustainable farming and fishing, of environmental education and of natural research. A necessary condition for moving in this direction, albeit insufficient, is the reduction of pollution in the lagoon environment. The reclamation work in the marsh areas which used to surround the lagoon and which acted as a filter for the hinterland areas, has facilitated the inflow of pollutants and nutrients from the drainage basin into the lagoon as well as making it more rapid. The drainage basin occupies an area of about two hundred thousand hectares and has a population of about 1,300,000 which includes the provinces of Venice, Padua and Treviso. Water borne pollutants reach the lagoon from this basin, carried by minor freshwater rivers and canals.

The load of general pollutants and of nutrients has reached a level which consistently exceeds the lagoon's self-purification capacity. In the past, this caused an increase in the likelihood of the phenomenon of macroalgal bloom with the possible negative effects of a loss of oxygen content in the water and of the emission of pollutants into the atmosphere. Furthermore, it increased the level of pollutants in muds and sediments. Other aspects pertinent to the environmental degradation of the lagoon are the existence of abandoned dumps and of high-risk industrial plants. The cumulative effects of past industrial waste disposal from Porto Marghera and solid urban waste also remain within the lagoon.

In the long term, the reduction in pressure on the lagoon environment resulting from anthropic activities will be tackled by making changes in technology and in the production structure of the activities whose emissions and waste are emptied into the lagoon and its drainage basin. Production industries which not only undertake production in the drainage basin area but especially in the Venice area, which is in closest contact with the lagoon environment, will be of particular importance. The development of the production structure and of the technology which characterises it is the essential core of the model of sustainable development for Venice.

2. Venice, the industrial society and the development of material production

Venice's entry into industrial society has produced a model of economic development from the second half of last century to the first half of this century characterised by material production as a priority. The crisis of the industrial model of the venetian economy was a crisis in the material production. In the 1980s there was an overall reduction in employment in the municipality of Venice; this employment loss can be almost entirely attributed to the material production sectors and, in particular, to the decline in the industrial centre of Porto Marghera. Industrial employment losses have also continued in recent years.

Nevertheless, material production is still of considerable importance in the Venice municipality: 30% of employed are still involved in material production activities. However, this is a notably lower percentage than that found at a regional level, where more than 50% of the workforce are involved in material production.

Material production in the municipal area is characterised by a dualistic structure: small- and medium-sized activities in both lagoon and mainland residential areas and large production units in Marghera. In addition, material production is mainly a characteristic of the mainland part of the municipal territory: 75% of those employed in material activities are on the mainland and only 25% in the historical centre and in the coastal and lagunar parts of the city. As far as production specialisation is concerned, in some material activities Venice provides a significant contribution to regional and national production, in particular, in refined petroleum products, in shipbuilding, in basic chemicals and in handmade and blown glass: apart from glass, all of these activities are located in Marghera.

High location costs make it unrealistic to predict an expansion of material production in the historical centre and in the coastal part of the city. Nevertheless, material production could play a role in the model of sustainable development in these parts of the city as well. The conditions for this occurring can be summarised in the consonance which the material activities demonstrate in relation to features of the environment and of local society which can act as real factors of location advantage.

An example could be the agricultural activities on the islands which can become high quality activities, even in ecological terms, and, as a result, target a refined group of consumers. An important component of material production, especially due to its direct connection with the lagoon's natural problems, is formed by the fishing sector. Lagoon fishing, which had experienced a contraction in favour of sea fishing, has expanded considerably in recent years due to the catch and earning possibilities permitted by new technology which does not require high investments in boats, but which has increased the risk of environmental damage. Fishing must be better linked with a requirement for sustainability which is defined in relation to the conditions of the ecosystem.

Another important sector of material production is constituted by minor shipbuilding and maintenance. In the shore areas and in historical and island Venice established firms exist which have managed to rationalise and overcome critical periods. Other enterprises also exist which can realistically set themselves market objectives related to high-income tourist and residential demand as well as to a probable demand for types of crafts adapted to the requirements for transport which respects the constraints relating to wave action.

Murano's blown and handmade glass industry is still important. In this district production techniques, unique in the world, and a highly qualified human capital make it possible to aim at the development of high quality output for a selected demand with high returns. A similar case can be made for lace-making on Burano and for a whole series of artisan activities of high production quality which are bound to the city's historical traditions and which can also aim at a growth guaranteed by a selected, quality-based demand. Examples of this are carpentry for boats and furnishings, furniture restoration, building restoration for those buildings with such special needs as those of Venice's historical centre have, the production of masks and the various forms of typical Venetian artisan production. Naturally, these are activities which must be exploited within the framework of market promotion and which, in any case, find a high level of integration with the city's specialisation in tourism.

The problem is to establish what the most appropriate measures of economic policy are for stimulating the maintenance and the development of this type of activity within a framework of economic sustainability and environmental compatibility. A constraint to every policy of this kind is that of not distorting conditions for competition and, therefore, of not creating conditions for market dominance and of barriers to entry which can arise from a protected and assisted development. On the one hand, therefore, one must concentrate on mitigating the conditions of differential cost which the location of production activities in the historical centre and coastal part of the city entails, mainly of transport costs. On the other, the focus must be on committing resources both to the training of human capital specialised in activities of typically local production and to creating conditions for the product's diffusion and its adequate commercialisation.

Yet the problem of material production is concentrated on the mainland and is, essentially, Marghera. In comparison to the 1960s, when the industrial area of Marghera reached its maximum level of employment, today the workforce has more than halved: the number of workers has been reduced from over 30,000 to 13,000. For each of the sectors which still continue to characterise this area, there are problems to resolve if it intends to be placed within a framework of sustainable development (Rispoli *et al.*, this volume).

Today Marghera is in a crucial transitional phase. The potential of activities connected to the post-industrial phase and to immaterial production are discernible, for example the activities which will develop around the project for the Science and Technology Park. However, factors of location advantage for the activities of industrial transformation still persist. In fact, this is an area which, already essentially separated from residential areas, can constitute a link between the sea and a wide market on the mainland as long as it can avail itself of an adequate system of infrastructures. Among the location factors at stake for the future of industry in Marghera, the environmental constraint plays a central role. There is no doubt that future investments cannot be made in plants at risk. They will have to have a minimum environmental impact and the areas available for location will have to be subjected to adequate environmental restoration work.

Conditions related to environmental impact and the level of risk to human health will play a particularly important role in the strategy relating to the chemical industry. In Marghera, this sector is faced with the need to make choices not necessarily bound by decisions relating to the growth or to the progressive elimination of oil traffic from the lagoon, since the chemical plants could also be supplied by pipeline. The significant decisions are related to the fact that the plants, despite the works which have been carried out and the alterations made, are, as a rule, old and high risk. In the current situation it is unlikely, not to say implausible, that the chemical industry's plants in Marghera satisfy the criteria established in European law on industrial risks. The alternative between decommissioning and radical (and obviously expensive) renewal is posited with increasing urgency. A decision seems to have been taken along the second line, that of restructuring with a highly acceptable level of environmental compatibility.

The provision of infrastructure, which above all concerns the transport system and accesses, is probably the most important problem to resolve in order to ascertain whether Marghera will be able to act as a link between the sea and the large production area of the Northeast; a role which many have predicted but which has never been fully achieved. Unfortunately, Marghera is a production area which would be improperly defined as coastal. Indeed, it is not directly located on the sea but on the lagoon; the real problem lies, therefore, in the future of the port. With a working commercial port, it is still possible to conceive that Marghera could have an industrial function which is characterised by converted and clean production. However, if the port functions fail, there will be no future for Marghera as an industrial area.

The structural composition of the port functions has changed radically over the last ten years and is destined to change again in the future. Future developments will move towards increasing both the importance of the commercial side and passengers. In spite of past resistance, which is marked by a somewhat ideal vision of public planning against the challenge of market trends, the model for locating port functions will follow the natural trend. This sees the commercial (and the remaining industrial) function located on the mainland's lagoon shore, while the port section for passengers is located in the historical centre.

The structure of the port has undergone considerable restructuring in recent years. Internal reorganisation and the port service's greater efficiency, in terms of unit cost of services supplied, are obviously indispensable conditions for guaranteeing the development of the port functions. However, they will not be enough unless the problem of infrastructure is tackled, and especially that concerning the sea. Clearly, the problem is one of thc passage of traffic into the lagoon. Furthermore, it is a problem which will also remain if and when the oil transport is completely excluded.

The problems lie in the canals and the accessibility for ships to the lagoon from the sea. The development of the commercial port will have to take the huge change in global commercial traffic into consideration. The principal feature of this change, which will mainly affects container transport, is the division of traffic into a world-wide network and a regional network. It is evident that Venice will have to compete with other ports, in the first place with those of the Adriatic, for transport of the second type from equipped platforms in the Mediterranean specialised in the first type. Indeed, the reference area is not only the Northeast of Italy, but Europe as well; it will not be easy to compete because, although it is true that Venice has a plain of enviable size behind it, it is also true that the port does not have the open sea in front of it, but the lagoon.

The experience with the *Canale dei Petroli* (oil-tanker channel) shows that the excavation of canals entails the risk of increasing erosion. Moreover, the possible increase in the frequency and level of high waters, within a global context of climate change which makes a rise in sea-level increasingly likely, raises the number of problems which must be solved in order to guarantee an adequate connection between the sea and the lagoon shoreline on which the commercial port is located.

Within the framework of sustainable development, the search for the conditions which render the development of the port's functions compatible with the survival of the lagoon ecosystem is a typical problem of cost-benefit analysis. As always happens in this type of analysis, the benefits of one choice are the costs of another. Among the benefits of developing the port, there must be included those which are typical of an infrastructure essential in the process of economic development, as is a commercial port: in particular, Venice's greater integration with the metropolitan area behind it and with the Northeast; the creation of value added and the function of stimulating production activity in the city and the region, exploiting the operations of production conversion on the lagoon shoreline and of other initiatives such as the Science and Technology Park. The total economic benefits should be able to contribute to covering the costs of maintaining the lagoon's hydro-geological equilibrium and the potential costs connected to a mitigation in restrictions for protecting Venice from high water in order to avoid jeopardising the port functions.

3. Towards sustainable tourism

Tourism is one of the economic activities which, at a global level, is recording the greatest growth both in terms of revenue and of employment even though, in many cases, it is an economic activity characterised by seasonal peaks and by a production structure based on unstable contractual relationships. The growing tourist demand can, in turn, be linked to the increase in per capita income and to the international expansion of economic growth.

An increasingly important part of tourist demand is directed towards the cities of art and the historical centres within wider urban areas. Venice, a city of art unique in the world, is the natural recipient of this growing tourism demand. Furthermore, it is a demand without limits if one considers that a visit to Venice is among the plans which every person in the world would like to carry out as soon as they have the financial possibility.

It is, therefore, completely natural that tourism is the economic activity which has had the most success in Venice and which continues to develop perpetually with the favourable drive of market mechanisms. Furthermore, it is also equally evident that its historical centre is the feature which attracts an increasing number of tourists to Venice. The main concerns are focused precisely on the effects of the pressure of tourism on the historical centre. In Venice's historical centre today there are more than 3 million residential tourists annually and about 6–7 million day-trippers.

Day-tripping visitors to the historical centre can be classified into three categories (Van der Borg and Russo, this volume): traditional day-trippers who visit Venice from their place of residence; indirect day-trippers who visit Venice leaving from another location which is the real destination of their trip; and false day-trippers or the commuter day-trippers, that is, tourists who, for economic reasons, stay in areas on the outskirts of the historical centre, even on the mainland in the same municipality of Venice, and restrict themselves to visiting the historical centre during the day.

The growth in the number of day-trippers belonging to the last two categories is a very clear indication of the expansion of the 'Venetian' area for accommodating tourists; i.e. of an area which allows commuter-trips to Venice's historical centre in only one day. This area already extends far beyond the boundaries of the municipality of Venice, along the littoral areas to the north and south of the lagoon above all.

The development of day-tripping tourism has already come to prevail over the development of residential tourism in the Venetian historical centre. This structural transformation in tourist demand directed towards Venice is the element which creates the most problems, both in relation to the city's economic development as well as in terms of policies for managing the tourist phenomenon.

The residential tourist has a greater impact in terms of spending than the day-tripper: the average daily spending in the historical centre of the former is almost double the latter's. In this sense, residential tourism is preferable to day-tripping. On the other hand, day-tripping is what can cause the most negative effects of congestion, environmental pressure and the wear and tear of monuments. For day-trippers, the costs of the city's use are potentially greater than the benefits in terms of income, since day-tripping generates earnings which generally go outside the area which produces them.

A typical problem of tourism's sustainable development is the identification of the threshold (carrying capacity) beyond which a further development of residential tourism in the historical centre can result in congestion and in a displacement of other economic and residential functions of the same historical centre. The limits to residential tourism's development formed by the scarcity of beds available in the historical centre (about 12,000 in comparison to over 90,000 in the entire municipal area) is only one of the constraints which act in determining the carrying capacity. Other constraints are constituted by the tolerance capacity of structures which would become subject to greater pressure as a result of an increase in residential tourism (typically, museums and churches).

A strategy of increasing the number of beds available for overnight-stay tourists in the historical centre must take the possibility of encountering one of these other constraints into consideration. From the social point of view, the most important of these is driving away the usual residents. In the current situation, it has been calculated that the other constraints make the optimal number of beds available lie between 14,000 and 15,000 (Van der Borg and Russo, this volume). An important element to take into consideration is the impact of an increase in passenger transport in terms of overnight stay tourism on the ships docked in Venice's port; the beds on berthed ships should be included in the number of beds available in the historical centre. This would also diminish, correspondingly, the need to adjust the existing availability of beds, if aiming to remain within the carrying capacity limit.

In every historical centre of art, there is a flow of tourists commuting from the peripheral areas of the city, as well as from neighbouring cities, and, often, it is precisely this flow which creates congestion problems. In this respect, the phenomenon of commuter day-tripping is an element common to all of the large urban areas in which an historical centre of art is found. In Venice the potential bottleneck formed by the limited routes of access could be the instrument to work on in order to manage such flows. Moreover, access routes are so well defined that they can be considered completely controllable. However, the historical centre of Venice is not, and is not intended to be, a mere destination for tourist visitors or for people involved in tourist activities. Indeed, it is and should remain the centre of a differentiated sum of economic activities and, therefore, a place of residence and access for all people who work in such activities. In these circumstances, the regulation of access for tourist flows cannot be handled and managed using the same procedures with which visitor access to a natural park is managed.

In the historical centre the indicators of the tourist supply's quality are not completely positive: on the one hand, the average price of a hotel room in Venice's historical centre is one and a half times (for the three star category) or two and a half times (for the four star category) the price on the mainland or in nearby cities such as Padua; on the other hand, the data shows that still a number of the hotels in the historical centre lack air conditioning, telephones, television and a mini-bar in the room. This can partly explain the results of a recent survey according to which, while more than 80% of day-trippers would return, only 40% of the overnight-stay tourists would return (Van der Borg and Russo, this volume).

In the Venetian historical centre, it is evident that other economic functions are being displaced by tourism and para-tourist activities; the latter concern commercial and restaurant activities' reorientation to tourism. The displacement is characterised by the fact that supply is structured so as to satisfy the growing amount of non-residential tourism, which is mainly a 'poor' tourism which has little care for and is insensitive to quality. There are indications that this change in supply in favour of day-tripping tourism is creating a vicious circle which progressively drives residential tourism out of the historical centre, especially due to the high discrepancy between price and quality of service given.

Alongside the gradual increase in the number of beds and in the resulting opportunities for staying overnight in the historical centre, there should be a strategy which gives priority in accessing the historical centre to day-trippers who come from the littoral and mainland areas of the municipal territory. In this way, the whole municipality would be involved in upgrading tourist activities in Venice. In any case, improving the quality of the existing accommodation structures still remains as a chief objective as well, particularly in the historical centre.

The unchecked development of a spontaneous scenario characterised by the expansion of mass tourism and day-tripping can also become unsustainable in the sense that, in the long-term, it seriously threatens the survival of the monumental, artistic and environmental heritage which forms the very basis for producing the tourist demand. This threat to sustainability, since it must be projected very far in the future, is not perceived in the preferences of the current consumers and not even in those of the current producers of the tourist supply. Perception of the threat requires a very low rate of time preference over a very long period which involves future generations and a much larger community, at a really global level.

A well-organised tourist demand and one which could reduce congestion to a minimum can be, on the other hand, one of the instrumental factors in a scenario of sustainable development. The capital which must be maintained in order to make economic activities, linked to satisfying the tourist demand, sustainable is the artistic and monumental capital, but also the environmental capital. The Venice Lagoon and its ecosystem can be a resource for sustainable tourism. There is room for promoting this type of tourism but, like the threats to sustainability which already exist in terms of maintaining the architectural and physical capital of the historical centre, similar threats can appear in the future when promoting and developing naturalistic tourism. A sustainable naturalistic tourism has to be combined with a regulation of access to the more ecologically fragile areas of the lagoon ecosystem and with action for environmental education. The Venice Lagoon could thus become the site of an experiment of international importance.

The scenario for a sustainable development of tourism requires a policy for transforming tourist activities into an economic base which contributes to financing the maintenance of Venice's infrastructure, monumental, artistic and environmental capital. This is an increasingly important objective since it becomes increasingly unlikely that the national and international community will make resources available for the maintenance of Venice's monuments and environment if they perceive that there is no corresponding contribution made from the earnings generated by the local exporting economic base; i.e. if they perceive that those who profit from this activity benefit from the results of efforts for maintaining the monumental, artistic and environmental capital without incurring any costs.

In this respect, it must also be taken into consideration that operators which are located outside the municipality, and even outside the region, enjoy the benefits of day-tripping to the historical centre of Venice. For these operators, Venice constitutes a positive externality because of the flow of income which it creates, but these operators are given no incentive to contribute to Venice's maintenance. Obviously, this is not a phenomenon typical of Venice; it is true for all Italian cities of art, even though to different degrees according to the importance of the artistic centre. It is a phenomenon which justifies a commitment to redistribute income gained as a result of the positive externality of being located in proximity to the city of art. The destination of this redistribution should be the maintenance of the heritage formed by the city.

In order to achieve the objectives of a sustainable tourism, various instruments can be used (Van der Borg and Russo, this volume). An important instrument is the regulation of flows. This is achieved through the introduction of accommodation bookings and car-parks at the terminals and through incentives which guarantee the visitor who books a series of facilities and discounts on urban services (multiservice card).

A second set of instruments tends to upgrade supply to meet standards of higher quality in such a way as to solicit a demand for it. Included in this set of instruments are those designed to improve accommodation provided by the hotel structures, to improve the quality of the museum supply, and to organise artistic events, catering for different types of cultural demand, by planning and spreading them over a broader period of time. To satisfy this cultural demand, Venice is already provided with an artistic capital which provides comparative advantages.

The new information and communication technology can play a crucial role in linking an improved tourist supply to demand so as to minimise the effects of congestion. The adoption of this technology should be encouraged in terms of installation and connection costs, as well as in terms of training of human capital adapted to using it.

4. Venice, a capital of the network of post-industrial society?

The development of the tourist industry constitutes the most evident and natural way in which Venice's entry into the post-industrial era is taking place. Tourism is a service, the growing demand for which is typical of post-industrial society and of the power of communication and movement which characterises it. Venice is given an incentive to specialise in tourism by market forces. The fact that these forces are so powerful and so apparently impossible to contrast means that there must exist a strong cultural attitude if one wants to make the orientation and control of these forces effective. The risk of *a priori* preclusion is that there will be an uncontrollable situation in practice.

The linking of tourism as closely as possible with other economic activities in the fields of artistic production, of conservation and exploitation of artistic and cultural goods, of environmental conservation, forms the basis of a policy which could tie tourism to the potential of the production and organisation of immaterial economic activities.

In contrast to the sector of material production, the immaterial economy and the tertiary sector in general do not show signs of an overall contraction in employment. The Venetian immaterial economy is showing signs of structural evolution towards the standardised model of immaterial economy in modern economic systems. However, in this process of transformation and adaptation, the Venetian immaterial economy is characterised by being too late in many respects in comparison to the evolution model of the immaterial economy in other centres of the metropolitan area of which Venice is a part (Rullani and Micelli, this volume).

This statement should, nevertheless, be qualified by considering the different structures and evolution of the immaterial economy in the historical centre and in Mestre. It is Mestre, rather than the Venetian historical centre, which exhibits a greater tendency to integrate with the model for the immaterial economy's evolution in the whole metropolitan area. This dualism also affects some activities which continue to locate in the historical centre but which are strongly attracted to the place where the economic and production trend is more dynamic; the headquarters of public administration and the university, for example. In conclusion, the historical centre tends to move away from the model which characterises the economic development of the metropolitan area and of the Northeast.

The battle which, since the second half of last century and, more evidently, in this century, Venice has fought for its inclusion in industrial society has brought about economic development and macroscopic imbalances and contradictions. The physical dualism of the city has been translated into an economic dualism (activities with a low material content in the historical centre and in the lagoon area, and heavy industry on the mainland shore). Moreover, a further economic dualism has been created between the Venetian mainland's specialisation in material production and the specialisation in material production in the Veneto region which surrounds it.

The result of these imbalances and contradictions has been the objective weakening of Venice's historical role as a capital city which it was assured by its cosmopolitan and inter-cultural dimension and by the use of the sea as a large communication network. For a certain period, from the 1920s to the 1950s, Venice's ruling economic class tried, initially with success, to transfer economic, entrepreneurial and financial resources into the venture for strengthening the connection between Venice and industrial society, availing themselves decisively of public support. In the post-war period, the plan for industrial expansion became a more explicitly public planning exercise but it came up against large changes in the international division of labour, against the crisis in public enterprise due to the increasing corruption on the part of political power, and against the emergence of a new environmental awareness.

The specific Venetian model entered into a crisis, but the small and medium-sized firms of the Veneto did not, and a series of fractures were created (one between the historical centre and the mainland on the one hand, one between the Marghera industrial area and the Veneto economy on the other) which broke up the large network and plunged the role of the city into crisis. It was no coincidence that in this period attempts were made to break up its administrative unity as well. The result was the 'passage from the long networks of the Venice capital to the short networks of Venice island' (Rullani and Micelli, this volume) and to accentuate the latter's tourist mono-specialisation at the same time as creating a crisis in the large functions of exchange and services of the former.

In post-industrial economies it is, by definition, at least theoretically possible to develop large networks of production and exchange without this also involving the movement of goods. The material content per unit of economic value produced is continuously reduced and there is a corresponding reduction in the content of use of the environment per unit of value produced. The immaterial economy is structurally in greater harmony with the environment compared to the industrial economy.

The cities become the favoured places for the growth of the immaterial economy. Indeed, the role of large cities, of capital cities, is closely connected to their capacity to play a role of leader in one or more functions which are typical of the post-industrial economy. There is no doubt that the growth of the immaterial economy offers Venice new opportunities for overcoming the imbalances and contradictions that have arisen in the recent past. These opportunities are especially offered to the historical centre which is provided with new means for escaping its inherently insular nature. In fact, the physical boundary formed by the lagoon is no longer as significant in terms of the criteria for communication networks in the immaterial economy. However, the error of thinking that only Venice's historical centre can be integrated into the network of the immaterial economy must be avoided. Indeed, if the mainland parts of the city can also hope to attract immaterial activities, the problem of their location advantage over the lagoon part of the city is posed anew.

The growth of immaterial specialisations, which appears as one of the potential features of the city and the urban area of the future, could occur in many cities. This will cause much competition between cities for securing the role of capitals in the new post-industrial economy. It must be noted that entire cities are in competition, not just parts of cities and, even less so, just historical centres. Venice offers reference material made up of cultural, artistic and environmental goods, which, in themselves, constitute a sum of location advantages for firms which operate in immaterial production. These are important factors in selecting locations but they are not so important as to conclude that this location will happen in the historical centre; it can easily take place near the historical centre, for example on the lagoon shoreline itself, if significant savings in cost are thus obtained in terms of mobility.

The expansion in Venice of immaterial production with a high innovation content and high value added depends on synergic action being taken by the firms which will potentially be able to choose to locate in Venice, and on the action of public policies which have to create the necessary context in terms of infrastructure and institutions, hopefully in co-operation with the firms. It is not easy to obtain all of the required conditions together. Historical experiences, beginning with the very famous one of Silicon valley (which everyone wants to, and precisely because of this cannot, imitate and repeat), show that a multiplicity of factors have to come together. These factors range from the availability of financial capital which is coupled with innovative ideas and entrepreneurial initiative, to the operational link with high-level research and training centres (it is no coincidence that two of the best universities in the world, Stanford and Berkeley, are found in Silicon Valley), to a cultural attitude characterised by the acceptance of business risk and of mobility.

A set of conditions are needed which might cause a critical mass of operators and firms in a whole series of advanced sectors of immaterial production to 'elect' Venice as the capital city in that type of activity. This demands the functioning of a combination of different services, from transport services, to those relating to residence, financial services, those of telecommunication, and services to the personal sector. Within a framework of co-operation with the firms and of project financing, political intervention should promote an adequate 'infrastructuralisation' of the city. This entails the provision of networks of infrastructure to allow the city to satisfy the requirements of developments in information and communication technology, but it also means that Venice will be able to supply services to the personal sector (education, health, culture) at required international efficiency standards.

Personal sector services will play a fundamental role in establishing a good relationship between the historical centre of Venice and the immaterial economy. It will be necessary to make the decision to reside in Venice on the basis of having a job in the immaterial economy, but in order for this to happen, the cost factors in the choice to reside in the city cannot be excessive in comparison to its qualitative benefits. Rich people who work in high value-added immaterial activities will be able to choose to come and live in Venice's historical centre even if this is relatively more expensive, but, together with the higher cost of this residence, there must be a correspondingly high quality of services provided as both support for carrying out production activities and for personal and family residential life.

The endeavour to attract activities involved in advanced immaterial production to the Venetian historical centre must overcome a further obstacle which tourism does not face. This derives from the fact that, while for tourism Venice's uniqueness can be made use of in an absolute way (every person in the world wants to and, at least up to a certain point, is willing to pay to visit Venice, which means that potential tourist demand is practically infinite), for the location of high value added immaterial production, this same uniqueness can only be exploited in a relative way. Moreover, a low quality tourist activity does not require particularly qualified human capital. This is a further disadvantage which a choice to set up an advanced immaterial activity in Venice has to overcome because it means that tourism is able to integrate more effectively with the features of the existing population and needs fewer adaptations in terms of mobility and changes in the qualifications of the resident population.

The spontaneous specialisation in tourism constitutes, in itself, a powerful incentive not to adapt the quality of services to the standards required by activities of high value added immaterial production or to the residents and those employed in such activities. Every strategy for promoting a high value added immaterial economic activity in Venice must allow for a further division within the immaterial economy between a part which lives on tourism and which will never be particularly concerned about the quality of services and another part which lives on high value added immaterial production and which is concerned about such quality. The spontaneous trends of the market tend to have the effect that the part bound to tourism becomes predominant with respect to the other and threatens to suffocate it.

This is why the upgrading of tourism is necessary, even if not sufficient, for the development of advanced immaterial production in the Venetian historical centre. Indeed, the two industries have an intrinsic correspondence and create mutual economies of scale. In any case, decisions to locate immaterial production on the mainland in the municipality of Venice should not be discouraged. Apart from anything else, it is very difficult to go against the location advantages created by the market without causing a widespread rejection on the part of businesses concerned. These are important stages in the construction of a specialisation plan in which the proximity to the artistic features of the historical centre and also to the environmental features of the lagoon will play an important role and into which the extended island and lagoon city can subsequently be integrated.

5. Sustainability and urban design

These considerations make it clear that the city's sustainable development requires an appropriate land use planning to support it. The application of the concept of sustainability at the level of the urban system causes value to be given to the diversity of functions as a condition for the system's capacity to self-regulate and maintain itself over time. The different areas of the territory considered should be set in a dynamic equilibrium by establishing a suitable system of relationships between them in order to avoid excessive specialisation and functional separation (Magnani, Pelzel, 1998).

The sustainability of the physical transformation process requires to take into account that what already exists exerts a determining influence by means of the inevitable effect of inertia. This inertia also relates to factors of the historical heritage which are to be given value in the sustainable urban plan. In the context of advanced societies in the process of moving from the industrial era to the post-industrial era, certain models of transformation in vogue in the past, and also recently, are bound to fail. For example, non-urbanised space no longer becomes a space awaiting urbanisation, but a territorial resource in itself, within a richer and qualitative vision of the anthropisation process. Similarly, urban growth is no longer synonymous with the expansion in housing and in physical infrastructure, but becomes, in itself, a concept which is qualitatively richer and which also integrates the functions of the non-urbanised spaces (Magnani and Pelzel, this volume).

In a similar new context, the number of projects on which it used to be possible to expect a broad consensus is becoming increasingly restricted because the conflicts of objectives over the use of the territory are more evident. The sustainable use of the territory is a recognised and accepted objective. However, as soon as it is made more specific, it reveals conflicts tied to the inherent deficiency of the territory as a resource.

The analysis of the projects for land use and urban design bear witness to the way in which Venice's entry into industrial society took place. In town-planning terms, this entry was seen with a shift in the centres of economic activity, and firstly of the port, towards the mainland's lagoon shoreline. This began at the end of last century with the construction of the Stazione Marittima. The shift of focus towards the mainland and towards industrial port activities was directed towards creating a large network of interaction, communication and exchange, to replace the historical network, which would be adapted to the needs of the industrial economy.

Crisis in this project has been confirmed, firstly by the large modifications on a global scale in the location of basic industrial activities, and then by the challenge created by post-industrial society for the cities. At the territorial level, the spontaneous evolution following this crisis has brought about a progressive separation in functional terms of the various parts of the city, to the point when the city's very unity is brought into question.

The need to respond to the opportunities for integrating Venice into the post-industrial era translates into the need to redesign the territory by taking a different scenario as a reference point as opposed to the spontaneous disintegration of the urban system.

As said, it proves futile to consider integrating the Venetian historical centre alone into the global network of the immaterial economy. The most appropriate reference point is the metropolitan area in which Venice is incorporated with its peculiarity in a process of giving value to environmental differences and specificity. The fundamental problem, which is however the most difficult to solve, is that of recovering the specific environmental peculiarity of Venice formed by its land-water relationship. In the past, Venice's relationship with the wider structure of the mainland was secured by hurdling one of the two components of the problem, i.e., the water, firstly with the railway bridge and then with the road bridge. In the future, this design could continue with a light railway which could have a sub-lagunar component.

Within the metropolitan area a sustainable urban scenario should be based on the strengthening of Venice's unitary functioning as a bipolar city in which the grounds for and incentives for integration take priority over centripetal ones. This is to prevent a potential concentration of projects on the mainland near the lagoon shore from continuing to keep the lagoon area in a state of marginalisation.

6. Actors and institutions in the Agenda 21 process for Venice

On the one hand, the complexity of the problem of Venice has given rise to a remarkable planning activity, even if it is all too often insufficiently defined for becoming operational in a satisfactory way. On the other hand, this complexity is seen in the form of a network of actors which is overcrowded and dispersed, in the sense that the problems of governance are confronted by a multiplicity of actors who, however, do not form relationships of mutual co-operation nearly as much as would be possible (Dente *et al.*, this volume).

The existence of composite structures for co-ordination (typically the Co-ordination and Control Committee that was set up with law 798/1984 and is known as the *Comitatone*) does not reduce the overcrowding of the network of actors nor make it denser. In situations of this kind, and therefore not only in the Venice case, in which formal participation is blended with a great deal of inexplicit pressure, the risk of overlapping powers of veto which increases the probability of stalling in decision-making is obvious. In this respect, the Venetian case represents, or has done until now, an example of the failure of an intervention model which is based on inter-institutional co-ordination from above.

From these points, two conclusions can be drawn on the institutional context required for a sustainable development of Venice (Dente *et al.*, this volume). The first is that it is necessary to avoid over-emphasising the speciality of the problem of Venice. If the problems under debate are perceived as problems only in so far as they arise in the Venetian context and in so far as they are specific to that context, it becomes increasingly difficult for a process of co-operation between actors to emerge; a process, on the other hand, necessary for achieving a shared and long-term vision to base programmes of action on. With a restricted and specific scope for defining the problems in question, it is more probable that each of the actors believes that their particular interests are being unilaterally sacrificed, in whichever comprehensive solution is proposed.

In reality, the factors of complexity in the Venetian problem arise from the impact made on the specific situation by the global transformations currently in progress. These range from more general changes, such as the passage from an industrial society to a post-industrial one, with the related consequences in terms of structure and methods of production and of the markets, to more specific changes, such as the effects of climate change, the new models of international seaborne transport and port organisation, the change related to the transformation of tourism into an activity of mass consumption, and the potential of new technology in immaterial production which is more beneficial to the environment.

Confronting the challenges posed by the epochal transformations makes it easier to escape the impasse of a zero-sum game, in which only a conflict of interests emerges, and to move on to a non-zero-sum game in which the possibility of mutual advantage in embracing the opportunities for change can be identified.

The second conclusion is of a more typically institutional nature. It results from the need, in order to become effective, for inter-institutional co-ordination to find its own legitimisation, not in an abstract plan which claims to be complete and rational, but in the proof of the emergence of social consensus which involves all of the actors who have an interest and an opinion to express on the problem of Venice. To this end, the responsibility for directing policies for the implementation of sustainable development should be assumed by the body responsible for institutional co-ordination (in this case, the *Comitatone*).

This should be achieved by giving a voice to all of the actors who have an interest in the way in which the problem of Venice's sustainable development will be tackled and resolved and who, therefore, have the right to be asked to assign a value, either for its use or its existence, to the 'Venice-good' and to the lagoon ecosystem in which it is incorporated. Among these actors, there is not only the local community but also national and international stakeholders since the sustainable development of Venice takes on the properties of a real 'global public-good'. Whether in terms of politics or of public opinion, the international community often makes its voice heard on the problems of Venice and its lagoon. This happens in an episodic way, on specific events, usually events which are considered dangerous and to be avoided. In any case, this is an important contribution because the expression of international public opinion helps local public opinion to avoid an egocentric position by perceiving the value which Venice has for the whole world.

Making the international community more aware of its positive and organic responsibilities in promoting sustainable development in Venice is one of the conditions required for implementing these objectives. This is not only for reasons of direct and indirect financial support, but also for reasons of a more fundamental nature which concern the very way in which the objective of sustainable development is defined.

7. Scenarios for sustainable development

Rullani and Micelli (this volume) propose various scenarios for the evolution of Venice. They construct the scenarios by accentuating some objectives, to some extent making them extreme, while sacrificing others. On the one hand, these extreme scenarios make it possible to identify the direction in which the Venetian system could go if some interests took priority over others. On the other hand, they make it possible to examine the implications in terms of risk for the realisation of a credible sustainable development.

The first scenario proposed is that of spontaneous growth. Essentially, this is the scenario of spontaneous evolution without significant public policy intervention. It is a scenario characterised, firstly, by the growth of a specialisation in tourism in line with the current model in which commuting is dominant, secondly, by a growing separation between the economic models of the historical centre and of the mainland and, as a result, by the progressive integration of the mainland alone into the metropolitan area. Finally, it is characterised by an unchecked evolution of Marghera, leading to its probable dismantling.

The scenario of spontaneous growth entails serious threats to sustainability. Indeed, it is a model marked by the risk of unrestrained pressure on the artistic, architectural and environmental heritage; by a progressive degradation in the city's social and cultural base with an increase in the likelihood of phenomena such as corruption and organised crime; by a widening gap between the short-term objectives of those who operate in the local economy and objectives for the long-term protection of the environment and the artistic and architectural heritage – a gap bound to create fewer incentives for the national and international community to take systematic action for this protection; by an increasing difficulty in organising a unitary administrative government of the various parts of the city and of its lagoon, with an inevitable increase in conflict.

A second scenario is the radical environmentalist scenario. This gives absolute priority to restoring the lagoon ecosystem to its state prior to Venice's industrial experience. Furthermore, it proposes intervention which is particularly drastic for anthropic ventures, such as the *Canale dei Petroli*, which are held to be responsible for exacerbating erosion. It is a vision ready to accept the elimination of port activity as a result of the fear that this can jeopardise the environmental equilibrium. It accepts (and desires) the exploitation of only those economic functions associated with a sustainable use of the lagoon environment. Moreover, it assumes an essentially planning-based and conservative approach towards tourism and the whole urban design.

Despite the fact that this is a model more concerned than others about protecting the environment, it entails equally serious threats to sustainable development. Indeed, it is a model which does not sufficiently exploit the opportunities for economic growth which are compatible with the conservation of the environment. As a result, this model would not have the strength to counteract the spontaneous forces of spontaneous development. The paradoxical outcome could be that of relegating the long-term environmental interests to a marginal weak position.

The third scenario is that of metropolitan integration. This is characterised by the predominant concern for integrating Venice into the metropolitan area and, thus, into the area of the Italian Northeast. The principal instruments for obtaining this objective are: maintaining and increasing the capacity of the port functions and of industrial activity on the lagoon shoreline; expanding the infrastructure for physical communication and accessibility for the entire municipal territory by means of new projects, such as a light railway; the future possible exploitation of Venice and its historical centre for executive functions.

The threats to sustainability presented by this kind of model for development arise from a vision which pays too little heed to the impact of the proposed solutions on the whole environmental system of the lagoon. There is also the danger of an excessive faith in the stage of constructing physical infrastructure and, therefore, of underestimating the prospects for development in the advanced immaterial economy. Finally, there is no guarantee of avoiding conflict with tourist specialisation in the historical centre.

Finally, there is the scenario of the accelerated transition to post-Fordism. This scenario concentrates on the conditions for achieving, especially in the Venetian historical centre, a development founded on the growth of high value added immaterial activities which is based on an insertion into global communication networks. This is the scenario which entails the fewest threats to sustainability; the danger is that it could be impossible to implement it. Indeed, it entails much uncertainty of economic success due to the strong competition which occurs at an international level. Moreover, such a model has to overcome conflict with the model of immaterial specialisation which is stimulated by the spontaneous market forces; that is, the model of a downgraded and unsustainable specialisation in tourism which characterises the scenario of 'business as usual' growth. For it to have any chance of success, it must avoid policies which, in the good intent of encouraging the establishment of immaterial activities in the historical centre, actually discourage entrepreneurs from choosing the entire area.

8. A sustainable development for Venice is possible

In reality, none of these extreme scenarios can hope to be a scenario of sustainable development because all of them share the threat to sustainability of being too focused on an over-specialised model. The diversity of urban functions constitutes an important element in a model for sustainable development because it increases capacity for reaction and for flexible adaptation to shocks. In this respect, some elements of a scenario for the sustainable development of Venice can be identified, combining features of the extreme scenarios just presented.

At the basis of every model for Venice's sustainable development is the concern for environmental preservation, understood as the preservation of the essential features of a lagoon ecosystem. Every economic development model has to establish a limit in its potential capacity of not jeopardising these features. Or rather, it must be able to generate enough resources to cover the costs of the intervention necessary for maintaining these features.

The model of development based on industry and particularly on the large basic industries located on the coast has no future. Certainly, there is room for some material production activities. It will be a question of managing the industries which remain in Marghera, especially the chemical industry, in a way compatible with the environment on one hand and with the future of the Italian chemical industry on the other. However, a development model which focuses on industry has no future.

There is no doubt that the great challenge of the future, formed by entry into the immaterial economy, opens up new prospects for economic growth which is compatible with the environment. The port activity itself, which is destined to accentuate its characteristics of passenger and commercial (container) transport, must be made compatible with this challenge in order to maintain its competitiveness which, in this field, signifies the chance for survival; this can be achieved by stressing the role of ICT in managing maritime traffic.

However, it must be clearly understood that in a context of the city competing fiercely for obtaining hegemony in the immaterial era, precise conditions must be created so that a critical mass of people and firms are led to choose to locate in Venice for immaterial production activities. These conditions concern the availability of infrastructure for networks and communication, of production and personal sector services, of training human capital and of the availability of financial capital. These circumstances cannot be obtained unless there is also a motivated entrepreneurial mentality and an efficient public structure. It must be recognised that it is much more difficult to create these conditions in the historical centre than on the mainland. As a result, a great effort must be made. However, it would be a mistake to focus on the entry of only the historical centre into the global network of the immaterial economy: prohibitive policies in this way could have counterproductive effects.

In any case, it is completely mistaken to believe that the development of the immaterial economy in Venice is an alternative to the development of tourism. The growing bias towards a tourism based on commuting and day-tripping poses serious threats to sustainability. The solution is certainly not to discourage tourism (which, anyhow, would not be possible, but which would create further bureaucratic complications and waste of resources). Instead, it is one of an upgrading of tourism in which the mainland part of the Venice municipality is also involved. This involves a series of measures: management of flows, the upgrading of accommodation, of the museum heritage and of the infrastructure for access; i.e., measures designed so that the tourist, or those who profit from tourism, contribute to the maintenance of the artistic and environmental capital which is the very source of income.

Sustainable development must have adequate institutional support. Given the importance of Venice and its lagoon as a global public-good, the actors involved are not only found at the local level, but also at the national and international levels. Venice and its lagoon are, on the other hand, a place where, both now and in the future, projects for environmental conservation and for the maintenance of the architectonic heritage must be carried out;

projects which require a huge mobilisation of resources which can be provided only at the national or international level. Therefore, there must be a body for institutional co-ordination which is able to give expression to the various opinions and to mobilise resources. The European community and the international community should have a place in this body. Efforts and resources should be committed to its efficient functioning. Indeed, if the institutional structure designed to take responsibility for such development is inefficient, all of the best projects for sustainable development will be bound to fail.

Bibliography

Costa P. (1993) *Venezia: economia e analisi urbana*. Milan: Etas Libri.
Dorigo W. (1972) *Una laguna di chiacchiere*. Venice: Emiliana.
UNESCO (1969) *Rapporto su Venezia*. Milano: Mondadori.
UNESCO (1995) *La Laguna di Venezia*. Venice.

CHAPTER 2

The Biodiversity in the Venice Lagoon as the Basis of a Sustainability Project

Mauro Bon, Danilo Mainardi, Luca Mizzan and Patrizia Torricelli

1. Sustainability as a cultural model for Venice and its lagoon

The idea of sustainability expressed as 'an improvement in the quality of human life within the limits of the carrying capacity of ecosystems' is undoubtedly the most culturally captivating challenge of the third millennium. Marking out sustainable routes for our way of life is a profound act of responsibility towards future generations which, however, implies a radical turn around in the relationship between man and nature up to now.

Man is a species which evolves culturally, and it is through his cultural evolution that he has been able to manipulate, for good and for bad, the physical, chemical and biological elements of the natural world for thousands of years. He is capable of influencing the evolution of living species by placing himself above or outside nature: nature in which he not only lives but in which he represents one of the components of the existing biodiversity (Mainardi, 1996).

The environmental state of emergency, of which all of humanity is now gaining awareness, comes from the lack of synchrony between the short time-frame of human cultural evolution and the longer time-frames for the biological evolution of the other species. The primary cause for alteration comes from this; if not through destruction of the natural balances, by the biological incapacity of other species, both plant and animal, to keep in evolutionary step with the rapid nature of change produced by man. In other words, the other species are incapable of developing suitable adaptive counter strategies.

The attempt to define new life strategies which are sustainable is no more than a logical consequence of the dramatic observation concerning the environmental damage produced by the excessive power which originates from

I. Musu (ed.), Sustainable Venice, 27–60.

human culture. Establishing a modern environmental culture must begin with a change in direction which leads from the current anthropic conception of nature to a biocentric attitude, by recovering the concept that man is part of nature. Furthermore, man is a conscious part of nature, and therefore one which must become active in order to recover the equilibriums lost.

The need for the integration of economic policy with environmental policy is repeatedly referred to in policy concerning development towards sustainability. In reality, the convergence between economics and ecology has been a critical point up to today. The two cultures, firmly founded on often antithetical presuppositions, find it hard to find a common language and, as a result, communication is still difficult.

It is without a doubt that economics is an anthropocentric science which pursues objectives that are incompatible on an ecological level. Such incompatibility is well expressed by Barde and Gerelli (1995): 'on a global level, growth remains the creed of humanity: to produce more, to consume more, to constantly increase welfare etc. This perpetual tension towards a growing exaggeration of products deriving from human activities does not exist in ecology and causes a profound incompatibility between the ecological tendency towards stability and the economic objectives of growth'. And not only this. Economics is a science which translates every phenomenon into monetary terms, and as a consequence, from a biological perspective, it furnishes analyses, estimates and predictions which are only valid in the very short term and are insignificant in evolutionary terms. It is the science which takes pollution into consideration only once it becomes a social cost and which values the decline of species in remote geographic areas as a phenomenon that does not enter into collective or individual utility functions.

Just from these few considerations, it is understandable that today, a cultural decision which seeks to combine economic growth with environmental quality is, in fact, a contradiction. Nevertheless, it is a contradiction which must be overcome in order to guarantee future generations the cultural foundations of sustainable development. An innovative culture, unrestrained by old conflicts between economic growth synonymous to environmental degradation and, in contrast, the search for environmental quality as a synonym of limitations. A new culture which knows how to translate ecology into a normative science in order to obtain a future founded, if not on biocentrism, at least on inspired anthropocentrism.

The economic structure and natural heritage of Venice and its lagoon make it a unique and exemplary case in which man, in playing the role of dominant species, can responsibly operate in the function of a shared vision of sustainable development. The natural heritage of Venice is composed of an entirety of diverse environmental typologies which is synthetically described in the following, highlighting the animal and plant communities which characterise it. Indicators elaborated on the basis of the community typifying each environment are then proposed. The phase in which the material subsequently presented was collected also stimulated some more general considerations. These can be summarised by the fact that a lack in 'environmental consciousness' exists among the various actors involved in the process for the sustainable future of Venice. The scarce awareness of the natural heritage, which is often simply considered as the pleasurable background colour against which the city is silhouetted, is the primary reason for the incapacity to realistically confront environmental issues.

To be more precise, numerous local, national and international institutions, which have scientific research as their statutory duty, look onto and work on the lagoon. However, there is little research which has environmental monitoring as its aim, either purely for the sake of knowledge or for implementing conservation projects or for a utilisation which could satisfy different levels of use. The lack of co-ordination between the various bodies in charge of research and the fact that the data produced is scarcely accessible is often complained about. This is confirmed by the data presented below, as its fragmented nature and frequent lack of methodological standardisation cannot be overlooked. Time series data on animal and plant communities, which could be used as reference for a basic comparison of before and after a certain event, do not exist.

On the other hand, within the variety of expertise present in scientific institutions in Venice, the figure of the biologist, or naturalist or ecologist is scarce and too often relegated to the role of collector of biological samples. Each environmental planning project involves the indispensable competence of engineers, architects and economists but not the equally indispensable presence of experts in biological matters. In fact, all too often, expertise of the chemical-physical kind, obviously useful for measuring the degradation and pollution, is exchanged for more general competence, thus neglecting the necessary attention to aspects related to biodiversity. The absence of a global approach has serious repercussions at the level of practical interventions, above all with regard to those for protection which tend to only be implemented in a state of emergency and with the token aim of treating the symptom. This is also confirmed by the legislation produced locally over time concerning protection issues. The deliberations on paper are abundant but Venice is certainly well below any quota for protected territory as suggested at the European level. It is enough to reflect on the fact that the Venice lagoon is presently not registered in the list of wetlands of international importance in accordance with the Ramsar Convention (with the exception of about 200 hectares of Valle d'Averto), while periodical monitoring results, regularly carried out only since 1993, can be seen to greatly exceed the criteria established by the Convention itself.

2. Natural aspects of the Venice lagoon

The following brief description of the Venice lagoon's different environmental typologies aims to furnish an analysis, though certainly not exhaustive, of the remarkable natural heritage which, along with its artistic heritage, makes Venice a unique and precious city.

In this analysis of natural resources, it is necessary to highlight the components as they are represented by the different species which characterise the various environments. Knowledge concerning plant and animal forms implies having a basic knowledge of their systematic positions, their exact scientific denominations (the common names of species vary from place to place), their ecological needs, the behavioural repertoire, an understanding of both the consistency and the state of health of the populations. This information makes up the entirety of basic knowledge which must be considered in any serious and informed project within the environmental sphere which has the conservation of biodiversity as its primary aim.

2.1. The coastal environments

The Venetian coastal area is divided into various environmental strips which are differentiated according to the diverse micro-climatic conditions resulting from the distance from the sea, differing wind exposure and the gradual enriching of sandy substrate with organic matter. Proceeding inland from the sea, the first environment is the waterline. This is characterised by tidal phenomena which accumulate a remarkable quantity of beached organisms. These then constitute the basis of an interesting trophic chain which attracts numerous species of birds of the order Charadriiformes, particularly during migration. An interesting endemic faunule survives among the waterline detritus, such as the beetles *Phaleria bimaculata adriatica* and *Xanthomus pallidus residuus*. Some species have recently become extinct due to the mechanical removal of the detritus (Bonometto, 1995).

Beyond a sandy strip which is bare of vegetation, there are the first plants which begin the important operation of colonisation and consolidation of the sandy shore. These are pioneer species with annual biological cycles which form the *Cakiletum* association. These plants have particular adaptations for retaining rainwater and avoiding dehydration, carrying out the important ecological role of trapping particles of sand with their root apparatus, thereby allowing successive species to take root. The amanthophilous plant communities are characterised by a marked capacity to adapt to extreme environmental conditions such as the high permeability of the terrain, the high salinity, the thermal excursions and the wind action which favours transpiration and evaporation of water.

A second plant association present is *Agropiretum*. In this environmental strip the vegetation, initially sporadic, becomes more variable The sand, held by the vegetation, accumulates more easily thereby forming the so-called embryo dunes, which enables the ammocolous *Ammophila arenaria*, a grass which creates real barriers for the sand pushed by the wind, to take root. Dunes form in proximity to the ammocolous vegetation and parallel to the coastline. These grow in height and protect the successive plant and animal communities from the wind (Bonometto, 1995).

Although the environments formed by both the *Cakiletum* and the ammocolous associations appear to be desolate, they host a rich and particular faunule. The gastropod *Theba pisana* has its northern limit of distribution here and the Italian wall lizard *Podarcis sicula* is found almost exclusively on the littoral strip. Among the avicolous species, two Charadriiformes should be noted, the kentish plover *Charadris alexandrinus* and the little tern *Sterna albifrons*, both of which nest in the few sandy strips untouched by seaside tourism.

The retrodunal environment, protected from winds and featuring a greater variety of vegetational types, has a rather complex morphology of higher ground and hollows. In the presence of stagnated water, the wetlands associations (*Schoenetum*), composed of mosses, rushes, sedges and canes grow, often with valences characteristic of the Upper Adriatic. The associations of the dry environments are represented by *Tortulo scabiosetum* and *Ammophiletum medicaginetosus*. The presence of alpine species, transported by rivers and established as a result of the presence of a fresh microclimate, are of great naturalistic and bio-geographical value (Caniglia, 1978; Gehu *et al.*, 1984; Pignatti, 1952–53;1959).

The retrodunal woodlands, once present all along the coastline, have today been substituted by pine forests made up of allochthonous species. Although these environments are important areas of refuge for fauna, they show an excess in woodland density and a flattening out of the variation in terrain, with scarce biological diversity as a consequence.

Today, the environmental elements described above are represented by a mosaic of few disjointed fragments which assume extraordinary value as a result of their testimonial nature. These areas are subject to pressure from seaside tourism and lack any type of management or control. Out of all the remaining biotopes, the only protected areas are the shores of Ca' Roman and the nature reserve at Bosco Nordio. Behind the old shore of St. Erasmo, the Bacan is an area of international interest in terms of fauna, where fruitful scientific research activities on the migratory avifauna have been carried out for years. Even in this case there is no plan for management and control despite the fact that the area is subject to intensive pressure from seaside tourism, fishing activities and the collection of molluscs.

In brief, the main factors of anthropic pressure acting on the shoreline are from the mechanical removal of waste and organic detritus which results in a notable impact on the faunule of the waterline, on the nesting of the avifauna and on the formation of embryo dunes. In the dune and retrodunal areas, already marked by uncontrolled construction for both private and tourism purposes and the use of soil for agriculture, pressure comes from the introduction of allocthonous species in the absence of woodland replanting activities, as well as the impact of certain sporting activities (cross country motorcycle racing, cross country car racing etc).

Indicators can be elaborated on the basis of the natural situation present and the existing knowledge of the morphology and the characteristics of the flora and fauna. The presence and height of the dune system , as is the depth of the shore, are for example, indicators of the various phenomena of erosion. An indicator of anthropisation is, in general, the presence/absence of the typical species and the autochthonous vegetation just described. More specifically for the retrodunal area, the presence/absence factors concerning the variation in the morphology of the terrain (dunes and depressions), the relationship between the presence of autochthonous and allochthonous arboreal-shrub species, the density of conifer populations etc, can also be taken into consideration.

2.2. *Salt marshes and shoals*

Going beyond the cordon of the shores broken up by the three inlets, on our route through the different lagoon environments we come across the water bodies in the basins and the emergent or partially submerged areas called *barene* (salt marshes) and *velme* (shoals). The salt marshes are tabular formations which generally take the shape of areas with slightly raised margins around internal depressions, crossed over by natural channels (*ghebi*). These often form a dense network which penetrates deeply and forms the pools called *chiari*. The salt marshes are usually characterised by extremely saline terrain and compact soils without porosity which are conditions that consequently block the movement of air in the deeper parts. It is, therefore, a strongly anaerobic environment, often with elevated levels of sulphur and hosting vegetation with a superficial root system. As a result of this, plants never reach significant heights. The salt marshes are colonised by diverse vegetative associations, often made up of monospecific populations, according to the different altitudinal and geographic zonation, and therefore, according to the salinity gradient of the environment. At the edges of the salt marshes, in the area most greatly subject to tidal action, we find a pioneer association, *Spartinetum*, almost exclusively made up of *Spartina maritima*, a grass which often brings about the elevation of the ground, thus leading to glasswort taking root and therefore the evolution of the shoals (areas normally submerged and only periodically uncovered during minimum tide) into salt marshes. The glasswort associations represent one of the most characteristic environments of the Venice lagoon; the most typical aspect is due to the presence of the endemic *Salicornia veneta* which often forms monophytic populations (*Salicornietum venetae*). They resist remarkable variations in salinity and preferably grow in the embryonal areas of the salt marshes or where the erosive phenomena are most intense. The halophilous vegetation of the salt marshes is more or less homogeneous. Existing differences are due to the salinity gradient and the diverse depth of the surface layer of soil; factors which condition the dominance of cover by one species rather than another. The associations already mentioned alternate with other halophytic associations which are mainly *Puccinellia palustris*, *Limonium serotinum*, *Arthrocnemum fruticosum*, *Aster tripolium*, *Salsola soda* etc. (Pignatti, 1996). In the less saline terrain, close to the brackish or freshwater, the typical halophytic species gradually give way to hygrophilous associations dominated by the presence of rushes, such as *Juncetum* and *Bolboschoenetum*, which are mainly found in the areas of the lagoon edge and the areas of the fish-farming valleys.

Among the few species of terrestrial invertebrates able to colonise these inhospitable environments in a stable way, some insects belonging to the Diptera (Empididae, Dolichopodidae and Ephydridae) and to the Coleoptera (Eteroceridae, Staphylinidae and Carabidae) can be found. These insects survive the periodic submersions by means of two different strategies. On the one hand, some species dig out tunnels in the compact soil of the salt marshes, inside of which, with the increase in water, a bubble of air is held thereby allowing the animals to survive. On the other hand, other species take themselves to the top of the vegetation with the rise in tide and come down again as the water goes down.

Less rare, even if never very numerous, are the salt marshes' typically marine species, in particular the Crustacea (Amphipodae, Talitridae and Isopodae) which find refuge from exposure during the low tides under the abundant organic material (prevalently algae) held together by the dense network of halophytic vegetation. Among the molluscs which are abundant among the roots of the vegetation and the relict material deposited by the tides, we find the Gastropoda Pulmonata *Ovatella myosotis* and *Auriculinella bidentata*. Inside the dense network of internal channels in the hollows and in the internal pools live *Truncatella subcylindrica*, a small Gastropoda Prosobranchia and *Hydrobia* (*Ventrosia*) *ventrosa* often together with the very similar *Hydrobia acuta*. Furthermore, some crawling benthonic species, capable of quick movements which allow them to follow the movements of the tide, thereby avoiding or reducing the exposure time, are found in the deepest parts of these depressions which remain covered by water during the most accentuated low tides. Among these we find the decapod crustaceans *Carcinus aestuarii* with the hermit crabs (Diogenidae and Paguridae) while *Upogebia pusilla* prefers to dig out tunnels inside which it hides from predators and possible periods of exposure. In addition, some fish move into the channels among the salt marshes, going up to the pool areas far within during high tide. Among these are some mullets, the flatfish *Platichthys flesus luscus* and some gobies. The populations of these very low areas situated in the lower levels of the meso-litoral plain, continue into the shoals situated at the same level and often at the salt marsh area perimeters.

At the top of the pyramid of the shoal and salt marsh environments, numerous aquatic bird species can be found throughout the year but particularly during the migratory and winter seasons (Table 1). Most of the aquatic birds which are dependent on the lagoon salt marsh complex are the so called mud-dwellers, belonging to the order Charadriiformes.

The salt marsh environment, subject to tides and without arboreal-shrub vegetation, is poorly adapted to nesting. Few taxa, generally only very specialised ones, nest in the most complete and stable salt marsh where they are rarely affected by high tide. Some species are considered rare both on the national level and also for the whole Mediterranean basin, such as, for example, the sandwich tern (*Sterna sandvicensis*). For this tern the Venice lagoon is the second most important Italian site and one of the few known in the Mediterranean (Scarton *et al.*, 1996). Indeed, the lagoon population of the redshank, *Tringa totanus*, makes up 95% of the entire nesting population in Italy (Scarton and Valle, in press). In total, there are about ten nesting Charadriiformes. In addition to the two species already mentioned, nesting of the following species has been confirmed: the oystercatcher *Haematopus ostralegus*, the avocet *Recurvirostra avosetta*, the black-winged stilt *Himantopus himantopus*, the kentish plover *Charadrius alexandrinus*, the yellow-legged gull *Larus cachinnans*, the black-headed gull *Larus ridibundus*, the Mediterranean gull *Larus melanocephalus*, the common tern *Sterna hirundo* and the little tern *Sterna albifrons* (Scarton *et al.*, in press; Valle *et al.*, 1994; 1995b) (Table 2).

Table 1 Over-wintering Charadriiformes in the Venice lagoon frequenting shoal and salt marsh areas, whose populations reach considerable proportions at the national level. Percentages are in relation to the total Italian over-wintering population in 1994 (INFS archive, from Stival, 1996)

Species	*%*
Grey plover	30.6
Dunlin	56.7
Ruff	21.2
Curlew	32.8
Redshank	22.4
Mediterranean gull	48.9

Table 2 Species of Charadriiformes nesting in the Venice lagoon salt marshes and number of couples counted. The data refers to 1996 and is partially drawn from Scarton and Valle (in press). The numbers marked with an asterisk are estimates

Species	*No. pairs*
Oystercatcher	1
Black-winged stilt	400*
Avocet	80*
Redshank	1000*
Mediterranean gull	1
Yellow-legged gull	3000*
Black-headed gull	88
Sandwich tern	385
Common tern	795
Little tern	50

The salt marshes and the processes of erosion: evolution over the last century
Until the restructuring of the lagoon inlets and the construction of the breakwaters at the mouths (Malamocco 1840–1872; Lido 1882–1910; Chioggia 1911–1930), the mouths were partially obstructed by sandbars lying perpendicularly to the coastline just beyond the shores. The ships, which were, however, relatively modest fishing ones, were therefore forced to carry out particular manoeuvres, such as coasting the beach at Lido or sometimes waiting for favourable tides in order to enter (see Fig. 1) (Favero *et al.*, 1988). With the construction of the breakwaters, the sandbars were demolished and the channel beds, subjected to the action of the tidal currents, spontaneously reached the depth of 9–10 m at Malamocco and 7–8 m at Lido. Around 1920, the Lido channel was excavated to the depth of 10–11 m. The salt marshes covered more than 20% of the lagoon basin surface area at the beginning of the century and were then reduced to 13% in 1930. Between 1919 and 1922 the Vittorio Emanuele III channel was excavated; this connects the Lido inlet via the Giudecca channel with the industrial area of Marghera. In the 1960s, a new and deeper channel, the *Canale dei Petroli* (the oil-tanker channel), was excavated in order to connect the new industrial area of Marghera with the Malamocco inlet. This cut the pre-existing Fisolo channel (which with a sinuous course ands numerous branches penetrated the central lagoon, thereby feeding the whole basin) with its rectilinear course up to the St. Leonardo port, then on to Marghera. The *Canale dei Petroli* was maintained for over 15 years at a depth of 12 m in the inner part of the lagoon and 15 m in the lagoon inlet areas. The entire remaining salt marshes equalled just over 8% of the lagoon surface area in 1970.

These erosive processes are accompanied by a reduction in lagoon surface area due to reclamation, dumps and closure of the fish farming valleys. From 1860 to 1896, over 4000 hectares were reclaimed for agricultural use while, since

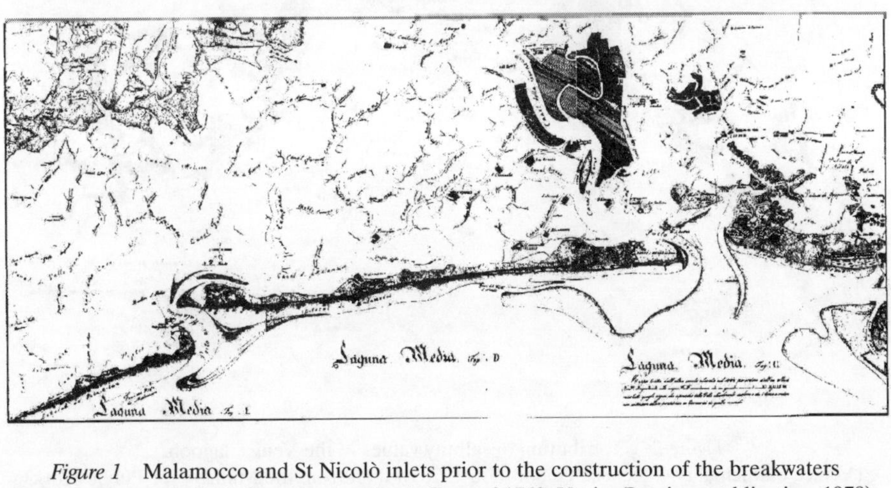

Figure 1 Malamocco and St Nicolò inlets prior to the construction of the breakwaters (from A. Bernardi, 1844, from the map by A. Emo of 1763, Venice Province publication, 1978).

the beginning of the century, 2200 hectares of the lagoon have been reclaimed for industrial areas and 7–8000 hectares for roads and services (eg. the airport). In addition, the closure of the embanked fish farming valleys deprived another 8500 hectares of lagoon from free water exchange. Therefore, from the middle of the nineteenth century until now, the lagoon basin has lost over 15,000 hectares in total, while the average depth has notably increased due to erosive processes which are caused by the navigation channels excavated by man. The lagoon has therefore become smaller and, at the same time, has become a deeper basin on average. Furthermore, it has channels which are more directly connected to the sea with greater capacities, thus accentuating the marine features to the detriment of those typical of a brackish environment with a high level of confinement, characteristic of an Upper Adriatic coastal lagoon.

These changes are highlighted by Figures 1 and 2. Figure 1 shows the Malamocco inlet and St. Nicolò in the nineteenth century situation prior to the construction of the breakwaters. The extensive sandbars in front of the lagoon inlet can be noted. Figure 2 shows the situation after the construction of the *Canale dei Petroli*. The average annual salinity values are highlighted in the different areas of the lagoon. It can be seen that in wide areas of the central and southern lagoon, the values do not go below 30‰, bar a few exceptions, and reach values close to marine ones (33–34‰). In fact, for wide areas of the lagoon salinity values coincide with the marine ones, thereby attributing very different characteristics to the basin from those pre-existing and typical of a lagoon; in some cases these values are actually greater than those in the marine

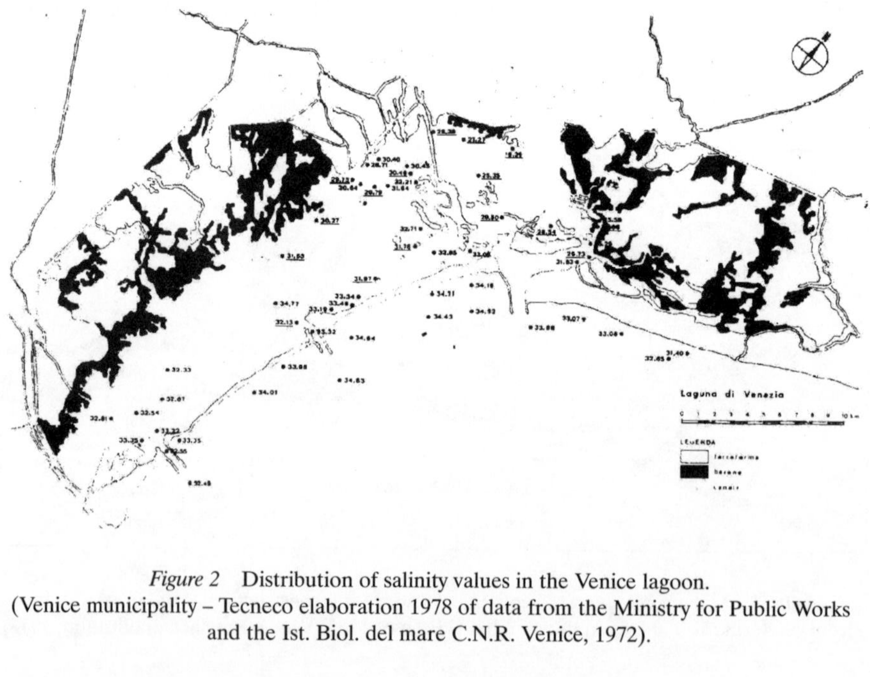

Figure 2 Distribution of salinity values in the Venice lagoon.
(Venice municipality – Tecneco elaboration 1978 of data from the Ministry for Public Works and the Ist. Biol. del mare C.N.R. Venice, 1972).

areas close to the estuarine deltas. Therefore, the imbalances and the alterations in the ecology of the lagoon ecosystem, induced by these environmental modifications, are easily detected.

The main problem for the salt marshes and shoals of the Venice lagoon seems to be represented presently by erosion which is largely a result of, directly or indirectly, human factors. In addition to the above-described alteration of the balance between the inflows of fluvial origin and the erosive action of the sea which has been caused by man, there are the fishing activities carried out with equipment of high physical and biological impact and excessive commercial and recreational traffic. Further erosive action is also caused by the channel excavations and other engineering works which cause modification in the lagoon hydrodynamics, thereby accentuating the processes of erosion and of transport of the fine fraction of the beds and, consequently, increasing the average depth.

2.3. The lagoon's aqueous environments

The extent of lagoon surface area covered by water varies notably throughout the phases of the tide. As it has already been seen, the salt marshes, which were once more widespread, are the generally emergent wide areas. The edges of these formations, together with areas of open lagoon, are positioned just under the average tide level and are therefore subject to periodic exposure during the low tides. These are called the shoals and the marshes respectively. The dense meadows of Zosteraceae, typical of these areas and particularly of the marshes, are also undergoing total reduction. The populations and species typical of these environments not only respond to the level of the tide but, moreover, to the degree of confinement of the areas in which they are located, maintaining, however, the same typology. According to the degree of marine action's direct influence on the various lagoon areas through the three Lido, Alberoni and Chioggia inlets, the basin is traditionally divided into live lagoon where tidal action is still directly felt, and dead lagoon, where the enlivening effect of the sea is felt in a less accentuated manner. Although this feature is prevalently related to the relative distance to the lagoon inlets, the extent and distribution of these areas does not follow a regular pattern through the basin. In addition to distance (as the crow flies), the channel's depth and course, the presence of islands or salt marsh systems as well as other factors which are able to limit or, on the contrary, favour the penetration of sea water during tidal exchanges, take on great importance.

The dead lagoon

The dead lagoon refers to all of the areas which experience the effects of the periodical action of sea water exchange in a marginal, or at least less accentuated way, during the tidal exchanges at the lagoon inlets. In addition to the distance from the inlets, the presence of channels and their depth, the presence of islands and salt marshes, marshes and shoals, continental water supplies etc, take on great importance in determining its extent. This type of environment includes the lagoon edges with valleys which are open or closed to varying degrees, the extensive marshes with shoals and salt marshes, which are particularly widespread in the northern lagoon, as well as the residual areas fed by freshwater which are often surrounded by reed beds in a variable way.

The selective factors for the populations are the extreme variability of the water conditions, in particular, the remarkable fluctuations in dissolved oxygen, salinity and temperature. In these areas, dissolved oxygen can go down to very low values during hot periods in which the water temperature can reach 30°C; values nearing those of anoxic conditions can be reached during nocturnal hours. The presence of silt with large quantities of organic substances caused by the inflow of continental water and low rates of renewal can lead to anoxic processes, and thus the decomposition of organic substances with the consequent production of toxic substances, particularly during the summer period. In the winter, the temperature of water with scarce movement and high surface area/depth ration can go down, as mentioned above, to surface values of less than 0°C. This sometimes causes the partial freezing of the surface in the lagoon shoreline areas. The salinity notably varies in relation to the inflow of continental waters, the influence of the live lagoon's tidal currents and the intensity of surface evaporation which is particularly acute during the summer periods. In this way, areas of very low salinity can be found close to freshwater inflows whilst the salinity can be near to, or actually greater than marine values in marshes or fish-farming valleys, which are particularly confined during the hottest summer periods.

This wide variability in water conditions is combined with that concerning the composition of the beds. Although these beds are composed of sands, silts and mud in a variable manner, they normally tend to have a greater abundance of finer fractions in the inner or confined areas (dead lagoon) and a gradually greater percentage of coarser components, mainly sandy, as progressing towards the lagoon inlets and the marine environment. The abundance of fine fractions (muds and clays) and organic substances in the dead lagoon make the penetration of oxygen in the lagoon bed difficult. As a consequence, the redox discontinuity layer gets nearer to the surface and can even reach a few millimetres in the inner areas of the lagoon.

These severe water and lagoon bed conditions select the species that make up the local populations and which, therefore, have to have either a wide ecological valencies or an extreme form of specialisation. In the areas bordering with the live lagoon or, at least, where extreme conditions are not reached, these populations are composed of the bivalve molluscs *Paphia aurea*, *Tapes decussatus* and the similar allochthonous species *Tapes philippinarum* and in the areas of lower salinity by *Scrobicularia plana*, by the gastropods *Nassarius reticulatus* and *Haminoea navicula*. In the most inner areas, which are characterised by a greater degree of confinement, there is the bivalve mollusc *Abra segmentum* (=*Abra ovata*) and *Cerastoderma glaucum* and gastropods

Hydrobia acuta and *Cyclope neritea*, the annelids *Hediste diversicolor*, the crustaceans *Corophium insidiosum*, *Lekanesphaera hoookeri*, *Cyathura carinata* and *Artemia salina* (Giordani Soika, 1949; Giordani Soika and Perin, 1974; Var. 1985;1986;1991).

It remains to be pointed out that in depth research in sample areas (Var., 1991) has highlighted how individual shoals or marshes can be complex environments, characterised by diverse levels of confinement (with relative variations of biotic community) in the more internal areas with respect to external or perimeter ones. As a result, in areas close to the lagoon inlets or the large channels which are characterised by low levels of confinement, populations typical of more internal areas and higher confinement levels can even be found in the inner parts of the extensive shoals which have few pools or high meadow coverage or macro-algae coverage in their perimeter areas.

The live lagoon

According to this concise subdivision traditionally used in the lagoon, live lagoon refers to all of the areas featuring a direct and notable influence of sea water. These areas, therefore, include the inlets and the areas surrounding them and extend to the inner parts of the lagoon in a diverse and irregular manner according to the courses of the main channels as well as to the presence of islands, salt marshes, marshes and lakes until bordering with the so-called dead lagoon. The easily perceived effects of the tide, which guarantee good communication with the sea, characterise the lagoon beds with a high presence of sandy components (reducing the mud and clay-silts fractions) and maintain the water at a good level of oxygenation as well as at a high and relatively constant salinity.

The populations present are marine, characterised mostly by a certain degree of euryhalinity and eurythermis and they vary in terms of the different taxonomic groups. While the planktonic populations are essentially marine (characterised by strong mortality when the planktonic stock is brought by the tides into the inner of the lagoon), the benthic ones have a gradient in relation to the extent to which the marine conditions are felt, going from the inlets to the inner lagoon.

The most common species are marine and thalassoid close to the inlet-channel areas as well as in the most immediate surrounding ones, such as the gastropod molluscs *Cerithium vulgatum, Nassarius reticulatus, Nassarius corniculus, Sphaeronassa mutabilis,* the bivalves *Solen marginatus, Loripes lacteus, Chamelea gallina, Mactra stultorum* (=*M. corallina*), *Mytilus galloprovincialis, Ostrea edulis, Crassotrea gigas, Donax trunculus*, and thalassoid with paralic inclusions in the gradually more inner parts (although still within the live lagoon) such as *Paphia aurea, Tapes decussatus* and *Tapes philippinarum, Loripes lacteus*. Analogously, among the annelids we find *Owenia fusiformis*, more typically marine, *Nephtys hombergi* and *Glycera tridactyla* (= *G. convoluta*); among the crustaceans: *Pilumnus hirtellus and Upogebia pusilla*.[1]

With regard to the plant component of the populations, the presence of marine phanerogams deserve particular analysis, both because of the importance of the biotic communities associated to them and the definite characterisation of the floor as determined by their distribution. These are superior (vascular) marine plants belonging to the class Monocotyledons, Order Helobiae which are equipped with a root apparatus that sinks into the lagoon bed and are able to consolidate the sediments on which they develop to different degrees. In dense populations, the luxuriant leaf system composed of nastriform fronds can constitute meadows which are sources of nutrition and refuge for numerous fish species, particularly during their larval and young stages (therefore they perform a nursery function). In the Venice lagoon three species of marine phanerogams are present: *Zostera noltii*, *Zostera marina* and *Cymodocea nodosa*.

Zostera noltii is the most widespread species in the lagoon. Due to its high environmental valence this species is able to colonise substrates positioned in the medium-low horizon of the meso-littoral strip, where it endures prolonged exposure during low tide, up to the typically infra-littoral beds. It prefers muddy substrates and colonises the areas near to the inlets and channels in both the more marine as well as the inner parts of the lagoon. It prefers, however, the areas of low turbidity and good exchange (Caniglia *et al.*, 1992a) (Fig. 3, Tables 3 and 4).

Zostera marina tends to colonise horizons which are notably lower than that of its similar *Z. noltii*, such as the marshes and also the edges of the salt marshes and channel beds. It also endures very low salinity levels (small nuclei

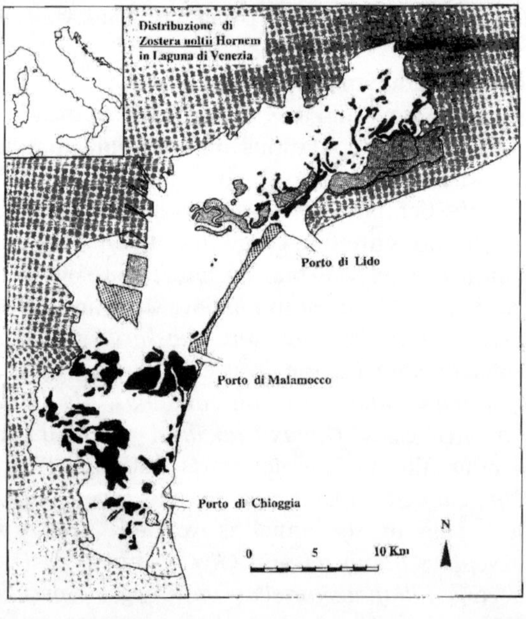

Figure 3 *Zostera noltii* distribution map. The main populations are indicated in black, the points refer to isolated areas (from Caniglia *et al.*, 1992a).

Table 3 Extent of the marine phanerogam populations in the Venice lagoon (from Caniglia *et al.*, 1992a)

	Pure populations (ha)	*Mixed populations* (ha)	*Punctiform populations* (number)
Zostera noltii	1430	2805	78
Zostera marina	240	3395	107
Cymodocea nodosa	390	1170	–

Table 4 Preferential environments of the three marine phanerogams considered (from Caniglia *et al.*, 1922a)

	Depth (cm)	*Substrate granulometry*
Zostera noltii	0–150	silty
Zostera marina	50–250	silty-sandy
Cymodocea nodosa	100–150	sandy

have been seen at the mouth of the Silone channel) but prefers beds which have lower levels of mud than that tolerated by *Z. noltii.* It is absent in the Central lagoon and has a punctiform and sporadic distribution in the Northern one while it forms dense and abundant populations in the Southern basin (Caniglia et al., 1992a) (Tables 3 and 4, Fig. 4).

Cymodocea nodosa is the least common and least widespread phanerogam in the lagoon. It is distributed, with the exception of some punctiform areas near to the St.Nicolo inlet, in the southern basin. It does not penetrate the inner areas of the lagoon basin, preferring areas with both good water exchange and sandy beds (Caniglia *et al*., 1992a) (Tables 3 and 4, Fig. 5).

Today, the phanerogam meadows, which are very important because of the positive influence they exercise on the environment (consolidation of lagoon beds) and for the direct effects on the biological populations (biotic communities linked to them), colonise wide areas of the lagoon basin by extending, according to the different species' different tolerances and preferences, into different lagoon environments. These populations, which are naturally subjected to a certain distributive variability and, perhaps, to a high natural turnover (Vatova, 1949), are particularly vulnerable to the aggressive action of human activities, in particular, to water pollution, an increase in turbidity, and above all, the use of fishing equipment which mechanically ploughs the lagoon floor (vibrating or non-vibrating mechanical scrapers, turbo suction, etc). It should be noted that, once damaged, a phanerogam meadow does not always seem able to spontaneously recover nor do areas formerly colonised always become

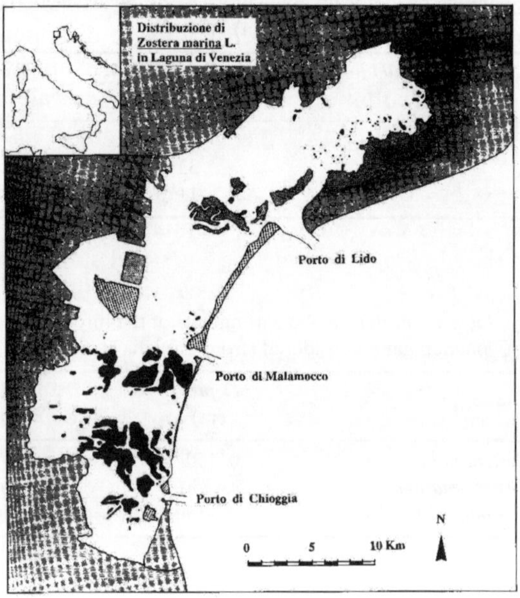

Figure 4 *Zostera marina* distribution map. The main populations are indicated in black, points refer to isolated areas (from Caniglia *et al.*, 1992a).

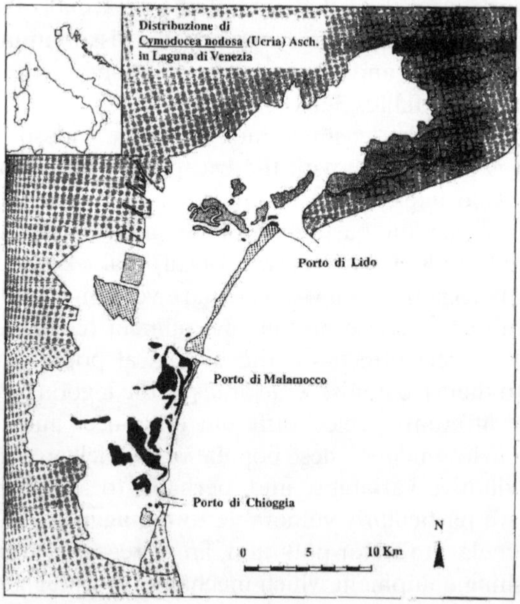

Figure 5 *Cymodocea nodosa* distribution map. The main populations are indicated in black, points refer to isolated areas (from Caniglia *et al.*, 1992a).

successively repopulated. It has recently been suggested that the cause of their drastic regression (and in wide areas, their complete disappearance; see Figs 1,2,3) from the central basin is not only due to clam fishing with mechanical scrapers, but also to water pollution of anthropic origins and the proliferation of macro-algae (this is in part due to hypertrophy of anthropic origins) (Sfriso *et al.*, 1990).

The state of the phanerogam meadows is certainly a valid candidate for being an indicator of the environmental compatibility of numerous human activities linked to the lagoon environment in general, and in particular, of the impact on the lagoon beds. Currently, there is up-to-date data available on the marine phanerogams in literature. On the basis of this knowledge, it is possible to develop, via constant systems of monitoring, indicators based on the relationship between the current extent of the meadows and the potentially colonisable surface area or that colonised in the past. In this way, time series of numerical indices which describe the state of the environment and furthermore, have predictive possibilities in terms of future evolutionary processes, could be produced. Moreover, the rarification of the populations and the general state of health of the meadows could provide effective indices of environmental risk. Finally, given the different ecological valences of the individual phanerogam species, the knowledge of their relative distributions can provide useful indications of the variation in the environmental conditions in terms of salinity, water turbidity, particle size of the lagoon bed etc.

The Casse di Colmata *(Reclaimed areas)*

The *Casse di colmata*, apparently unrelated to the environmental context of the open lagoon, are areas reclaimed at the beginning of the 1960s by means of filling in wide areas of salt marshes. These were supposed to constitute the third industrial area in Venice though this work was subsequently abandoned (May 1969). Following the abandonment of the project, the reclaimed areas were recolonised by rich vegetation and its related fauna, thereby forming an area which is remarkably valuable in terms of its fauna.

In particular, the most interesting biotopes are the areas B and D-E which are areas of 410 hectares and 780 hectares respectively. The ground morphology is rather heterogeneous, both because of the soil characteristics (clays and sands) and its elevations which go from 0 to 2 m a.s.l.. As a consequence, the environments present vary greatly in terms of the presence of halophyte and hydric vegetation, wide freshwater channels, channels and little woodlands on the higher knolls. (Calzavara, 1979; Caniglia and Salviato, 1983).

In terms of fauna, the reclaimed areas represent an enormous storage tank for the invertebrates (Canestrelli, 1979; Ratti, 1979) but, especially, for the extraordinary quantity of aquatic birds present throughout the seasons of the year. Among the numerous nesting species, the most important are the shelduck, the avocet, the black-winged stilt, the marsh harrier and many others. Over-wintering and migratory species are also numerous (Rallo, 1978; Stival, 1996; Valle *et al.*, 1995a).

2.5. *The fish-farming valleys* (Table 5)

The natural landscape of the fish-farming valleys, which is more lagoon-like than any of the others, is the result of centuries of interaction between man and nature. In its most traditional form, a fish-farming valley can be defined as a 'historicised' environment of great natural and anthropic importance, in addition to its economic importance due to the fish-farming which traditionally takes place there.

Table 5 Fish-farming valleys of the Venice lagoon. (ASAP Fish-farming Valleys Cartography Source): Current situation, surface areas and species raised

Valley	*Total surface area* (ha)	*Aqueous surface area* (ha)	*Fish species raised/method*
Southern Lagoon:			
Brenta	–	–	(inactive)
Millecampi	2041	1608	(inactive)
Morosina	563	400	mul.; e/bas.; e/gil.; eel;
Ghebo Storto	228	150	mul.; e/bas.; eel;
Pierimpiè	528	440	mul.; e/bas.; e/gil.; eel;
Rivola or Barenon	–	–	(inactive)
Figheri	420	300	e/bas.; e/gil.; eel;
Zappa	380	350	mul.; e/bas.; e/gil.; eel; pra.;
Cornio Alto	110	90	mul.; e/bas.; e/gil.; eel;
Cornio Basso	200	170	mul.; e/bas.; e/gil.; eel; biv.;
Marina Averto	294	254	mul.; e/bas.; e/gil.; eel;
Contarina	353	300	mul.; eel;
Serraglia	400	300	mul.; e/bas.; e/gil.; eel;
Northern Lagoon			
De Riva	300	150	mul.; e/bas.; eel;
La Cura	40	10	(inactive)
Santa Cristina	42	12	(inactive)
Ca'Zane	150	75	mul.; biv.;
Dogà	1650	1000	mul.; e/bas.; e/gil.; i/bas.; eel; stu.; biv.;
Grassabò	860	750	mul.; e/bas.; e/gil.; i/bas.; eel; pra.; biv.
Dragoiesolo	1080	980	mul.; e/bas.; e/gil.; eel;
Fosse	172	160	mul.; i/bas.; i/gil.; eel; pra.;
Lio Maggiore	27	16	mul.; e/bas.; e/gil.; eel;
Cavallino and Basegia	508	400	mul.; e/bas.; e/gil.; eel;
La Bianca	62	55	mul.; i/bas.; i/gil.; eel;
Capanno	87	63	mul.; e/bas.; e/gil.; eel;
Falconera	40	31	mul.; e/bas.; e/gil.;
Liona	150	27	(inactive)
Olivera	65	50	(inactive)
Paleazza	536	333	mul.; e/bas.; e/gil.; eel;
Musestre	–	–	(inactive)
Sacchetta and Sacchettina	115	74	mul.; e/bas.; e/gil.; eel; biv.;
Sparesera	20	14	pra.;
Saccagnana	27	22	(inactive)

mul. = mullets; bas. = sea bass; gil. = gilthead; eel = eel; pra. = prawn; biv. = bivalves; stu. = sturgeon; i/ = intensive; e/ = extensive.

The majority of the fish-farming valleys are found in the most inner reaches of the lagoon next to its edges. They are wide pools of water which were originally confined by salt marsh systems to form basins in constant communication with the lagoon. These have been gradually isolated from the original lagoon environment via barriers which have become increasingly less precarious and permeable to marine action until becoming the present definitive earthen banks.

In total, the valleys represent a wide surface area and constitute the area of interface between the mainland environment and the lagoon/marine one. As a result, since their formation they have taken on a very important role as a stage in the trophic migration processes of the different fish species which periodically penetrate the lagoon from the sea, moving inside the lagoon basins in search of food. In fact, in the favourable seasons these basins have great feeding resources due to the striking loads of nutrients of continental origin, which are, moreover, optimal conditionals for the growth of many euryhaline fish species which enter in their juvenile stages, to then return to the sea at the beginning of the winter season.

Until a few decades ago, at the beginning of spring, the so called phenomenon of 'the ascent' (*monta*), was conspicuous and consistent. This is the penetration of the valley waters by the fry of the mullets *Mugil cephalus, Liza ramada, Liza aurata, Chelon labrosus* and *Liza saliens*, of the gilthead *Sparus aurata*, of the sea bass *Dicentrarchus labrax* and, during different periods, of the eel *Anguilla anguilla* and the flounder *Platichthys flesus luscus*.

Since ancient times,[2] the populations of the area have taken advantage of these migratory phenomena. In particular, of that which corresponds to the autumnal 'descent' (the *smontada*), which is the period of return to the sea via the same route travelled in spring. Here, the fish were easily captured by forcing them through tortuous waterways culminating in a trap. The valleys were banked by *serrage*, barriers made of marsh cane, which allowed the passage of fish in the post-larval and juvenile stages during the ascent while impeding the fish which had become adults from going out again except via the appropriate routes used to aid capture during the *fraima* (the local term used to indicate the migratory movement of the fish towards the sea).

This system of natural fish-farming, which was based on natural processes by taking advantage of the instinctive migratory behaviour of the fish, also maintained the valleys 'open' i.e., in constant equilibrium with the lagoon water level.

However, over the last few years, water pollution and the drastic reduction in the quantity of young fish in the period of ascent, has caused the valley fish-farmers to completely close the basins with fixed banks (earthen banks) which have, in fact, transformed the valleys into closed basins which are maintained in communication with lagoon water by closure systems (*chiaviche*) employed as desired by the designated person. Freshwater is supplied through other closure mechanisms and also through artesian wells (also in order to raise the temperature of the water during winter, given that the groundwater is constantly a few degrees above zero). As a result, the valleys have become areas which are no longer directly subject to the expansion of the tide, and the process of supplying juveniles of the fish species raised has to be carried out by taking the fry from natural environments. Consequently, there are new types of fishermen, the 'fry-fishers' (*novellanti*) specialised in the capture of large quantities of young fish to then be sold to the valley farmers.[3] Today, the fry-fishers, who were the sustenance of valley fish-farming until very few years ago, have nearly disappeared. The reasons are multiple: the increasing scarcity of natural young fish has rendered this type of fishing unprofitable, thereby progressively driving the fishermen's activities to different, more guaranteed ones. Today, many young fish for the valley businesses are, essentially, imported. Finally, the development of artificial reproduction techniques for the main species raised has relegated the supply of natural young fish to a secondary role.

It is known that the primary productivity of a coastal lagoon system is the highest existing among natural systems. Comparing the data for net biomass production (m^2/year or Kcal/m^2/year) of an environment such as the Venice lagoon with others, also terrestrial ones, it can be seen that this reaches levels higher than values of forest terrain or even artificially fertilised agricultural ones (high technology intensive agriculture) (*var*., 1985; Bianchi, 1988). A rational and compatible use of the fish-farming valleys with the lagoon environment could therefore be a useful direction for the development of the local fishing sector's production.

Although, the productivity of the fish-farming valley environments is, in theory, very high, it is not used in a very efficient way given the extensive system of farming. In this system, once the fish is put in, it is almost totally entrusted to the environmental resources, with very limited intervention from the fish-farmer. If, on the one hand, this farming technique produces low fishing returns per unit of surface area, on the other, it reduces the farming management costs, at least in terms of input derived from farming intervention. Instead, the cost of maintaining the banks and the cleaning of the basins, activities which continually employ people, remain a consistent part of the overall costs. However, with respect to the more profitable and efficient intensive systems, such farming techniques have, without a doubt, the advantage of a smaller environmental impact which, in some cases, is virtually nothing or, indeed positive for some aspects (in particular in terms of fauna). The pollution generated by fish-farming is often negligible, as it involves processes with minimum human input (just feed and chemical substances) and wide pools of water to which public access is impossible, thereby offering particularly receptive areas for numerous animal species. Among these are many birds.

Nevertheless, fish production which is essentially destined for the local market, is viewed positively in terms of quality, thus being preferred to imported production. However, the quantity produced is largely insufficient for internal (local) demand and the imported products are sometimes sold as local produce or from valleys in order to increase consumer appreciation. The market niche would, therefore, be receptive to greater production, above all, for valued species such as eel, penaeid shrimp and sea bass, which are hindered by technical and farming problems. In particular, eel farming (which was once flourishing and subsequently abandoned for other species with shorter life cycles, more easily supplied and considered more profitable) is currently being re-evaluated. The scarcity of the product has rendered the albeit low national demand greater than the supply, driving the product to prices which make it highly profitable. In this case, there remains the serious problem of the total reduction in eel fry which are increasingly difficult to obtain (Table 6). However, from the employment perspective, the number of workers who find employment in valley fish-farming is limited, given that the total number of employed in the sector is 150 people (Boatto and De Francesco, 1994) (Table 7). On calculating the surface area of the basins used in active valley fish-farming businesses to be about 6855 hectares, the value indicated, of 1–2 employees for every 100 hectares of surface area used for these practices, is plausible. This highlights how this productive situation, potentially important from the environmental point of view, also due to its low impact on the surrounding systems, is in reality, very inefficient from the productive perspective and almost negligible from the perspective of employment.

Table 6 Total production data, per hectare and as a percentage, for the species raised. (From Boatto and De Francesco, 1994)

Species raised	*Total quintals*	*Quintals* (ha)	*% of the total*
Eel	791	0.13	15.11
Penaeid shrimp	25	–	0.48
Sea bass	800	0.13	15.28
Gilthead	1272	0.21	24.30
Mullets	2317	0.38	44.26
Carp	30	–	0.57
Total	5235	0.85	100.00

Table 7 Permanent and temporary employment, and family members employed in a valley fish-farming (From Boatto and De Francesco, 1994)

Permanent employees	
Valleys employing them (%)	87.50
Total number	79.00
Average number	3.29
Temporary employees	
Valleys employing them (%)	50.00
Average number	143.33
Family Members	
Valleys employing them (%)	4.17
Total number	3.00
Average numbers	0.16

It remains to be stressed that the advantage of the low environmental impact is reduced or cancelled out by the increasingly prolonged closure of the fish-farming environments from exchange with lagoon water, which is justified by the farmers in terms of the increase in the pollution of the lagoon. The, by now almost permanent, closure of these basins to these tidal exchanges not only impedes the well known ideal (given its already illustrated rarification) phenomenon of ascent of the young fish but also notably reduces the volume of the lagoon basin with the hypothetical (and at the same time contested) effects on the phenomena of high water, which it is said to magnify.

The submerged environment

The valley lagoon beds are normally composed of medium-fine sediments, with the rare exceptions of those situated in areas corresponding to ancient shore areas where a sandy fraction can be found (Barillari and Rosso, 1975; Hieke-Merlin *et al.*, 1979).

The benthic populations (autochthonous) are made up of euryhaline species, well adapted to rich fine floors of organic material. The population compositions are slightly diversified in relation to local conditions; primarily to salinity. Therefore, we can consider the lagoon valleys as mesohaline (5–20 S%), polihaline (20–30 S‰) and hyperhaline (>30 S‰) with notably different values of productivity and biomass. For example, the biomass of benthic populations can vary respectively, from 79 to 108 g/m^2 for the mesohaline (the least productive), 208–282 g/m^2 for the hyperhaline, to 339–368 g/m^2 for the polihaline (the most productive). The majority (from 80 to over 95% of total weight) of the zoobenthos is formed by the biotic communities *Cerastodema glaucum*, *Loripes lacteus* and *Abra segmentum* (=*Abra ovata*) (Vatova, 1981). In fact, the populations appear to be characterised by low specific diversity and high biomass values in accordance with the environmental characteristics which prove to be particularly selective for the organisms. Also present, although representing lower fractions of the biomass, are the crustaceans *Idotea baltica, Gammarus* sp., *Sphaeroma serratum* and *Cyathura carinata*, along with the common crab *Carcinus aestuarii* (Vatova, 1981). Other molluscs which manage to colonise these environments in a stable way are the gastropods *Haminoea navicula, Cyclope neritea* and *Bittium reticulatum*.[4] The only Echidnoderm which enters the fishing valleys appears to be *Amphipolis squamata* which seems to be in accordance, together with the characteristics of the populations just described, with what is foreseen by the theory of paralic domain (Guerlorget and Perthuisot, 1983). In environments with lower renewal rates (such as lagoons or coastal ponds), there are distinct environments or influence areas with features which are strictly marine as opposed to continental ones, thus subdividing them into confinement areas. Relatively homogeneous groups of populations would correspond to these different areas of confinement rather than to the absolute characteristics of the local chemical and physical parameters (principally salinity, oxygenation and temperature of the water).

With the closure of and the progressive increase in the isolation of the lagoon valleys, (due to the increased pollution of the lagoon water, many valley managers have greatly reduced the opening periods of the closure systems connecting with the lagoon, above all, in the summer) the phenomenon of confinement in these areas could be further accentuated.

The emergent environments

As highlighted above, the variations in the salinity in the embanked fish-farming valleys do not follow the regular rhythm of the tides or natural river inflows. The closure of the valleys has therefore led to the characterisation of some biotopes (particularly in the strip along the lagoon shoreline) with a greater presence of freshwater or weakly marine environments. A sharp reduction in typical salt marsh vegetation results from this while the shoals tend to be eliminated since they are not considered useful either as emergent or submerged environments. The reduction in the shoals, moreover, means the loss of the embryonic conditions which lead to the formation of new salt marshes. Instead, the presence of embankments as well as elevated and consolidated islands leads to the development of arboreal-shrub type of vegetation, similar to that of the hinterland.

From the natural perspective, the most valued areas are probably the environments characterised by the presence of freshwater which can be considered as substitutes for the biotopes present in the past, when greater interaction between the mainland, fluvial areas and the lagoon existed.

In spite of the fact that these areas have been and continue to be modified by human intervention (and are, therefore, not characterised by high naturalness), they are important as a result of not being subject to frequent human use. The modification of the landscape into one with a scarcity of shoals and salt marshes and an abundance of pools and riparian environments is an advantage to swimming birds and a disadvantage to the mud dwellers. Indeed, as a whole, the lagoon valley area can be considered as one of the most important wetlands in Italy for the over-wintering of Anatidae and of the Coot. It should be noted that most of these species are nearly exclusively found inside the fish-farming valleys despite the fact that their surface area represents just 18% of the total lagoon surface area. However, an important factor in the distribution of some species is the different hunting pressure inside of the fish-farming valleys with respect to that in the tidal lagoon. Nearly all of the valleys are in fact, are subject to the Fauna-hunting Institute's regulations and have hunting density indices which are notably lower than those referring to the remaining lagoon surface.

The presence of barely accessible suitable areas has always been an important factor in the nesting of the Ardeidae, colonial animals which often establish themselves with polyspecific populations in both environments of pure reed beds and along the embankments furnished with arboreal-shrub cover. Nesting, particularly with regard to the red heron (*Ardea purpurea*) populations, has shown a great increase over the last decade. The Venice lagoon represents the most important nesting area in Italy and an area of international importance for this species. Other significant data concern the first recorded nesting of the great white egret *Egretta alba* in the Veneto (the second in Italy) and the first reproduction of the grey heron *Ardea cinerea* in the Venice lagoon in this century (Table 8) (Amato *et al.*, 1994). The lagoon fish-farming valleys are also the over-wintering sites for some of these species. For example, the over-wintering population of the night heron *Nycticorax nycticorax*, with over 50% of the national population, prove to be the most important in Italy (Stival, 1996). The over-wintering populations of the little egret, grey heron and great white egret are also important at the national level.

2.6. *The lagoon shoreline environments*

In its original conditions, the Venetian mainland environment was characterised, in rapid succession, by salt marshes, freshwater marshes, riparian woodlands and mature woodlands, all intersected by resurgent rivers and flanked by swampy clearings. Today, in a territorial system which has completely mutated and fragmented since ancient times, there are natural peculiarities remaining in only a few areas which, due to their relict conditions, take on extraordinary importance. These are, in particular, small fluvial and woodland biotopes as well as disused quarries abandoned by man and re-colonised by nature.

The areas characterised by the presence of weakly saline or fresh water are found at the lagoon edges, near to the few rivers and channels which flow into the lagoon and into some of the fish-farming valleys where the salinity of the water is varied artificially by means of the closure systems. These are environments which are quite widespread throughout the territory and which, although limited in terms of their extent, are important as genetic tanks for

Table 8 Number of nests found in the two heronries of the Venice lagoon. P = non quantified presence, A = absent (from Amato *et al.*, 1994)

	1991	*1993*	*1994*	*1995*
Valle Dogà				
Little Egret	P	80	P	A
Purple heron	81	214	230	355
Valle Figheri				
Night heron	P	P	133	99
Squacco heron	A	A	5	8
Little egret	737	193	550	608
Great white egret	A	1	A	A
Grey heron	A	4	26	52
Purple heron	206	187	219	324

plant and animal species that were once common in the lagoon and mainland interface areas.

The algae and floating vegetation ascribed to the *Charatea* and *Lemnetea* associations can be found in proximity to the stagnant water which is mainly found in the old quarries in the municipalities bordering Venice. Instead, the vegetation forms real submerged meadows of *Potamogeton* and *Ceratophyllum* in the resurgent river courses with clear and more oxygenated water. In the fluvial bends, the marshes and, in general, where the water is slower, plants with leaves laid out on the surface of the water are rooted to the bottom (*Nymphaea alba* and *Nuphar lutea*) (Anoé and Caniglia, 1987; Caniglia *et al.*, 1992b). Among the most widespread associations is the phragmites–marshes, almost exclusively made up of the marsh reed *Phragmites australis*. These are typical transitional vegetation and are abundant in all of the areas of inflow in the fish-farming valleys and in the marshes, also the saltwater ones. In the areas where only freshwater is found, it can be accompanied or substituted by another cane *Typha angustifolia* and *T. latifolia* or, in the marshes which are being filled in by different types of sedges. Hydric scrub grows along the edges of the wetlands and most frequently on the banks. This has prevalently arboreal and shrub elements typical of riparian woodlands (*Salix alba, S. cinerea, S. purpurea, Alnus glutinosa, Ulmus minor, Populus alba* and many other species).

The perifluvial inland environments and the old quarries assume great importance as genetic tanks for small fauna. Here, research has identified numerous new species for the Veneto, for Italy and for science (Raffone, 1991; Ratti *et al.*, 1995). In addition, the presence of amphibians and reptiles is interesting (Borgoni *et al.*, 1994). These areas are often within anthropised contexts and play important roles as refuges for many migratory bird species. Among the most important nesting species are the little bittern *Ixobrichus minutus*, the little ringed plover *Charadrius dubius*, the lapwing *Vanellus vanellus*, the kingfisher *Alcedo atthis*, the penduline tit *Remiz pendulinus* as well as numerous warbler species (Amato and Semenzato, 1988).

Finally, in terms of fauna, all of the areas of the fluvial inflows with reed beds are of considerable importance. In addition to the above-mentioned Ardeidae, there are other important groups of Anatidae, Rallidae and Passeriiformes nesting in the fish-farming valley areas. A typical bird of prey in the valley and shoreline areas is the marsh harrier, *Circus aeruginosus*, which nests inside the reed beds. Among the most interesting small mammals are Miller's water shrew *Neomys anomalus*, and the harvest mouse *Micromys minutus* (Bon *et al.*, 1993).

In the Venetian shoreline area, only one relict plain woodland remains as testimony to the original primary forests. This is the whole of the Carpenedo woods and the park of the Villa Matter (Mestre). The Carpenedo woods is a relict of plain forest which can be ascribed to the *Querco-Carpinetum boreoitalicum* association. The peculiarity of the vegetation in this environment has been extensively shown by Caniglia (1981). This is a site of exceptional importance from the didactic and scientific point of view since it has elements of flora and fauna from the primary forest which was once widespread throughout the territory and today has disappeared. The woodland vegetation is composed of a dominant association of *Carpinus betulus* and *Quercus robur* as well as a rich nemoral florule among which are *Anenome nemorosa*, *Vinca minor*, *Polygonatum multiflorum*, etc.

Another valuable environment is found outside the woods and at the adjacent Fort Carpenedo: these are the so-called wetland meadows, made up of uncultivated flat land with seasonal water pools in which interesting relict associations (*Molinietum*) are found together with rather interesting flora (Bon *et al.*, 1996; Caniglia, 1981). The presence of certain invertebrates in the Carpenedo woods is significant (Cesari and Orlandini, 1984; Ratti, 1984): the perimeter ditches host notable species such as the bivalves *Pisidium obtusale* and *P. personatum* and the beetle *Donacia polita* (only Venetian site). Other relict elements which were linked to the, once widespread, forest environments are some of the Carabidae beetles (*Notiophilus rufipes, Metallina lampros*) and Cerambycidae, as well as some surface dwelling molluscs (*Helix pomatia, Perforatella incarnata, Tandonia budapestensis*) which are currently only frequently found in the alpine and pre-alpine biotopes. The birdlife includes numerous woodland species among which are the tawny owl *Strix aluco* and the great spotted woodpecker *Picoides major* among the nesting species, the woodcock *Scolopax rusticola*, the green woodpecker *Picus viridis* and the sparrowhawk *Accipiter nisus* among the overwintering species, as well as numerous perching birds (Roccaforte *et al.*, 1994). Among the small vertebrates should be noted the relict populations of Italian agile frog *Rana latastei*, a Padana plain endemism and finally, the hazel dormouse *Muscardinus avellanarius* which, among the rodents is considered a good indicator of integral woodland ecotones (Bon *et al.*, 1996; Richard and Semenzato, 1988).

The dimensions of this environment also represent its current limits: weak micro-climatic variations can in fact cause the extinction of the most sensitive species. For example, the lowering of the groundwater aquifer level as a consequence of the nearby drainage channel excavations has brought about modifications to the hydrological characteristics of the terrain with related consequences on the flora and fauna. Actions which are apparently of no influence, such as taking away organic material from the beds, can provoke alterations in the soil-type structure of the terrain. Not to mention the illicit hunting activities or the collection of edible invertebrates and mushrooms or the trampling resulting from the uncontrolled visiting of the area.

On the other hand, the easy location of these few fragments, small woods, quarries and marshes within a completely urbanised landscape, makes them evident and therefore more easily controlled in comparison to the extensive lagoon environments. The fact remains, however, that these are without any real protection and any form of nature management, with the exception of the Gaggio quarry, managed by LIPU and the 'little lakes' of Martellago which have been defined by the Municipality as a nature reserve of local interest for the last few years.

3. The Venice lagoon in the overall picture of internationally important wetlands

In considering the subject of wetlands the Ramsar Convention, also signed by Italy in 1976, considers aquatic birds to be among the best indicators of environmental quality[5] and of the richness and diversity of communities. Within the context of the Ramsar Convention, criteria regarding the presence of aquatic birds have been fixed. These establish whether a wetland can be defined as being of international importance, a fact which then obliges the committed State to guarantee adequate measures of protection and management for these wetlands which take on international importance. Two objective criteria establish that an area is of international importance. To be recognised as such, it must:

- regularly host at least 20 000 aquatic species, or
- host at least 1% of the individuals of a world population of any aquatic bird, in any period of its annual biological cycle.

Regular censuses of the water birds wintering in the wetlands in the Venice province have been conducted since 1993. These follow the census methods standardised at the international level, as indicated by the International Wildfowl and Wetlands Research Bureau and the *Istituto Nazionale per la Fauna Selvatica*. The winter period is considered best for estimating the populations of aquatic birdlife in the Paleartic Region. During over-wintering, the birds are strongly bound to the territory and do not make significant migratory movements. Moreover, many species (waterfowl, waders, etc) tend to gather together in their search for food while others (herons, cormorants, etc) collect together in nocturnal roosts: allowing for a more effective count. Therefore, constant monitoring of the water bird through the use of standardised methods enables population trends to be checked and important assessments on the quality of the host area to be carried out. Naturally, in order to reach these results, many years of sampling and a correct evaluation of data is necessary. The analysis of this data should also consider factors from outside the sampled areas (meteorological and climatic conditions at the European level, rate of increase or decrease of nesting population etc).

If a first qualitative–quantitative indication can be drawn from the study of the population trends, another type of objective value is given by the criteria suggested by the Ramsar Convention. So far, the censuses of the Venice lagoon have revealed the presence of 74,642 individuals in January 1993, 100,048 individuals in January 1994, 120,925 individuals in January 1995 and 122,775 in January 1996, with an average of 104,598 individuals (Baccetti *et al.*, in press). In every census the results have greatly exceeded the number of 20,000 aquatic birds as stated by the Ramsar Convention. The criteria of 1%, stated in order to single out wetlands of international importance, is based on estimates of the total of an aquatic bird population which is determined periodically. The most recent values published have been taken into consideration in applying this criteria to the case of the Venice lagoon (Rose and Scott, 1994). The species which have exceeded the threshold value of 1% in at least one of the four years of censuses are listed in Table 9. Among these, there are at least five which have exceeded the threshold value in at least three cases. These could be considered indicative in relation to the Ramsar Convention: black-necked grebe *Podiceps nigricollis*, great white egret *Ardea alba*, shoveler *Anas clypeata*, coot *Fulica atra* and dunlin *Calidris alpina*.

Therefore the Venice lagoon exceeds the threshold values set out by the Ramsar Convention for at least 6 parameters (more than 20,000 over-wintering aquatic birds, 5 species above the 1% level), whereas exceeding just one threshold value justifies inclusion in the wetlands considered in the list of areas of international importance. In spite of this, it is currently not on the list of wetlands of international importance with the exception of about the 200 hectares of the WWF fauna refuge at Valle dell'Averto.

3.1. Nesting birdlife as an indicator of an area's natural value

There has never been a complete census of the nesting species of the lagoon's aquatic birdlife carried out. The problems involved in this type of research are numerous and are, above all, the result of the absence of a body to promote and co-ordinate this activity. There are problems of logistics, such as the difficulty in gaining access to areas which are privately managed, in particular, to the fish-farming valleys. Furthermore, there are technical problems: many species which tend to be gregarious in winter tend to disperse during nesting; this is true, for example, for the waterfowl which is a group for which no quantitative data is available.

There is, however, much data concerning specific groups, such as for the nesting salt marsh species (terns, gulls, redshank, etc). In these cases, the censuses which exist have been carried out for some years (Scarton *et al*., 1997), but, as

Table 9 Species exceeding the 1% criterion in at least one of the four censuses carried out in the Venice lagoon. Species usually exceeding such criterion are indicated in bold (from Baccetti *et al*., in press)

Species	*Level 1%*	*1993*	*1994*	*1995*	*1996*	*No. years*
Black-necked grebe	**1.000**	**1.833**	**3.552**	**3.619**	**3.205**	**4**
Little egret	800	612	1.522	1.205	663	2
Great white egret	**120**	**262**	**152**	**335**	**265**	**4**
Shelduck	150	69	10	300	466	2
Wigeon	6.000	1.241	2.712	7.712	4.439	1
Teal	10.000	4.497	5.216	10.259	12.272	2
Pintail	3.000	764	1.256	4.508	4.270	2
Shoveler	**1.800**	**1.163**	**2.488**	**3.836**	**4.373**	**3**
Goldeneye	200	96	295	239	109	2
Coot	**20.000**	**28.880**	**32.957**	**31.472**	**37.164**	**4**
Dunlin	**14.000**	**10.800**	**16.644**	**22.310**	**17.385**	**3**

a whole, the research is still at an embryonic stage and prevalently regards the distribution of the species throughout the territory (cf Bon *et al.*, in press).

The ecological value which the nesting of a species assumes is very high; the species/environment relationship is even more important here than for over-wintering since the territory is not only chosen for the presence of food, but also because it offers a secure and peaceful environment in which to bring up offspring. The use of nesting species as indicators of environmental quality has, in fact, been used and perfected over the years, even in environmental impact assessment projects (Bogliani, 1990). The difficulty of giving an objective value remains, although only recently, some authors have attempted to do this. In addition they have succeeded in defining numerical values and a sort of 'classification of importance' for the nesting species in Italy (Brichetti and Gariboldi, 1992). Nevertheless, the application of these methodologies still remains difficult given the diversity which exists between the individual environmental situations in our country.

In the absence of numerical series and objective values, it must be borne in mind that even the sole presence of a nesting species (or even better, of a population) is already an important item of data. In order to have a good indicator, we will follow the general rule of using the species which are very demanding from an ecological point of view due to their strict relationship with a particular nesting territory. Typical examples are the species which nest in environments in delicate equilibrium (coastal areas and salt marshes), in selective environments (reed beds) or in ecologically mature environments (plain wodlands). The example which we will examine is that of the coastal area, in which two well known species, the kentish plover (*Charadrius alexandrinus*) and the little tern (*Sterna albifrons*) nest.

An example qualitative indicator for the Venetian coastal areas: the kentish plover and the little tern

The kentish plover is a species which is widespread throughout the coastal areas of the northern Adriatic which have not been transformed for tourism. More than 140 pairs were found on the Venetian coastline in 1992 (Cherubini and Panzarin, 1994).

The Venetian population assumes certain importance, above all, when compared to the figures available which indicate that there are 1500–2000 nesting pairs in the entire Italian territory. The coastal area of Ca' Roman is the most important reproductive site in the Venetian shoreline, with about 50 nesting pairs every year. The majority of nests are dug out in covered or hidden areas, often under the leaves of the *Xantium italicum* or at the base of the clumps of *Ammophila littoralis*. The kentish plover chicks leave the nest the first day after hatching and follow their parents along the beach in search of food. The first flight occurs about 30 days after hatching.

The little tern is the second species considered as an indicator. As it is very demanding from an ecological perspective, the little tern also nests along the Venetian shoreline but it is more localised than the previous species. It is a migrating species which arrives in Italy in April and at the end of the month, it identifies the site where the colony is formed. In the second third of May, the first depositions occur. The little tern nests in colonies and only uses a small part of the shore. About 6,000 couples of little tern, which represents about 30% of the European population, nested in Italy in 1983 and 1984. In the same year, more than 500 pairs were surveyed in the Venice lagoon, although recently the Venetian population seems to have undergone a marked reduction. Ca' Roman is, in fact, the only Venetian shoreline to host a stable colony of little tern. On other shores, (Punta Sabbioni, St. Nicolò del Lido etc) the attempts at little tern nesting have often turned out to be sterile due to disturbance by bathers. In contrast to the kentish plover nests, the little tern nests are constructed far from vegetation, on the detritus strip which is deposited during the winter tides. Generally, three eggs are laid which are then incubated for about 22 days. The chicks leave the nest a few days after hatching and hide in the vegetation where they are fed small fish by their parents. They begin to fly after about 20 days but continue to be followed and fed by the adults for some weeks.

The kentish plover and little tern eggs, which are deposited directly onto the sand, have a high probability of being destroyed. The principal risk is that of having the nests crushed involuntarily by bathers who visit the coast. The extremely mimetic colouring of the eggs is, in fact, useful for reducing predation from other animals but it is counterproductive in the case of an environment visited by tourists. In spite of surveillance, even at Ca' Roman, the Kentish Plover and Little Tern continue to lose their nests in rather high proportions. The causes cannot only be attributed to 'natural' events which are, in some way, part of the balance of reproductive strategies, but also to predators which are foreign to the local fauna: the rats and feral cats; the former are favoured by the environmental degradation, the latter by their uncontrolled release.

Therefore, the nesting of kentish plover and little tern can be used as an indicator of anthropic disturbance. These species are sensitive to seaside tourism (trampling), to the presence of tourism structures, to the mechanical treatment of the shore and to predation by species related to anthropisation of the territory. In this case, the presence/absence factor can be used as a simple qualitative indicator.

Other species that lend themselves to this type of use are, for example, some of the salt marsh nesting birds, which are particularly sensitive to the edaphic and vegetation components of the salt marshes and anthropic disturbance. As a result of these characteristics, they nest exclusively in some little disturbed and undamaged areas of the Venice lagoon. The most important are the black-winged stilt, the avocet, the redshank, the common tern, and the sandwich tern. This way of thinking could be extended to other selective environments such as the reed beds. The indicators to consider are species which are almost exclusively from this environment, which nest in some fish-farming valleys or in the areas of fluvial inflow, for example, the marsh harrier, the purple heron, the bearded tit and some warblers.

Conclusions

The future prospects for the natural heritage of Venice must begin with serious and conscious action in terms of protecting both the ecological biodiversity, which is meant as the variety of environmental typologies, as well as the biodiversity of animal and plant species which have evolved adaptive mechanisms in order to live in these environments.

Conserving biodiversity means sustaining the entire ecosystem, by maintaining the useful and unrepeatable combinations of genetic information kept in the various organisms which then reach us by means of millions of years of evolution.

A scenario of compatibility for Venice and its lagoon can only be elaborated by starting from this premise. Any project for the real sustainable development of Venice must, therefore, be developed with the far-sightedness necessary to maintain and value the natural and cultural diversity of the city over time.

In the local debate on the future of Venice, an antithesis often emerges. This focuses on how an environment which would not exist without the constant protective works of man, can be defined as natural. Therefore, the lagoon is considered as an artificial environment and so as exclusively employable, exploitable and able to be manipulated by man.

It is our opinion that the discrimination between natural/artificial is both extremely difficult to define (is man not part of nature?) and, nevertheless, of very little use. What emerges as important is the capacity to make choices which have, in the long time-frame of biological evolution, a value for the survival of our species and of the other species in the ecosystem which provides for them.

Notes

1. It should be noted here that a lot of species present in the lagoon have a wide ecological valence and therefore do not have a precise significance, since they can live both in the sea and in the various lagoon zones. Although not listed in the text, since they do not determine zonation, they can be abundant and sometime constitute a considerable fraction of the existing biomass. Among them, for example, the crustacean *Carcinus aestuarii* (= *C. mediterraneus*), some isopods and Gammaridean amphipods, the bivalve mollusc *Scapharca inaequivalvis*, the gastropod *Cyclope neritea* etc.

   Furthermore, many species, although they have preferential habitats, can also be found in considerably different habitats, with different frequency and sometimes also size. For example, the bivalve *Cerastoderma glaucum*, although being a typical paralic species, can also be found in the *baccani* of the S.Nicolò inlet-canal, intermingled with the typically marine species *Chamelea gallina*. Likewise, in a similar habitat, near the Malamocco inlet-canal, specimens of *Venus verrucosa*, a typically marine species are found (however, almost never of medium-large size), intermingled with *Paphia aurea* and *Scapharca inaequivalvis*.
2. The valleys were probably already used in pre-roman times. Fishing systems based on fish migrations (*piscariae acquae*) began to be mentioned around the year 1000, while the first reference which corresponds to the current concept of 'valley' dates back to 5th July 1425 (in *Rogadis* deliberation of the Venice Republic's Senate in which *clausura valium* is mentioned).
3. The character of the fry-fishers is quite a controversial one, even among fishermen who accuse them of causing a considerable reduction in the catch in impoverishing natural fish stocks by strongly interfering with juvenile stages.

4. It is probably *Bittium scabrum*.
5. Aquatic birds are among the most representative species of wetland biotic communities and, for this reason, they are widely used as ecological indicators. The term 'aquatic birds' usually refers to avifauna which is ecologically dependant upon wetlands.

Bibliography

Amato, S. and Semenzato, M. (1988) Sull'avifauna di alcune cave dell'entroterra veneziano: 3,. *Lavori Soc. Ven. Sci. Nat*., **13**, 115–133.

Amato, S., Semenzato, M., Borgoni, N., Richard, J. and Tiloca, G. (1994) Status attuale delle popolazioni di ardeidi nidificanti nella laguna di Venezia (Italia N-E). *Riv. Ital. Ornitol*., **63**, 200–204.

Anoe, N. and Caniglia, G. (1987) La vegetazione acquatica e palustre di alcune cave di argilla dell'entroterra veneziano. *Lavori Soc. Ven. Sci. Nat*., **12**, 159–175.

Baccetti, N., Bon, M., Cherubini, G., Semenzato, M. and Serra, L. (in press) La laguna di Venezia: zona umida di importanza internazionale per lo svernamento degli uccelli acquatici, *XIII Convegno del Gruppo di Ecologia di Base G. Gadio: Aspetti ecologici e naturalistici dei sistemi lagunari e costieri*. Venice, 25–27 May 1996.

Barde, J.T. and Gerelli, E. (1995) *Economia e politica dell'ambiente*, Bologna: Il Mulino.

Barillari, A. and Rosso, A. (1975) Prime notizie sulla distribuzione dei sedimenti superficiali del bacino settentrionale della laguna Veneta. *Mem. Biogeogr. Adriatica*. Suppl., **9**, 3–32.

Bianchi, C.N. (1988) Tipologia ecologica delle laguna costiere italiane. In: *Le lagune costiere: ricerca e gestione*.CLEM, Massa Lubrense (Napoli), pp. 57–66.

Boatto, V. and De Francesco, E. (1994) *L'economia ittica in provincia di Venezia: dalla produzione al consumo*, Osservatorio Economico del Settore Ittico, A.S.A.P. Azienda Sviluppo Acquacoltura e Pesca, Venice.

Bogliani, G. (1990) Elementi di attribuzione di valore a specie ed aree nella VIA. *Agricoltura*, **204/205**, 92–105.

Bon, M., Borgoni, N., Richard, J. and Semenzato, M. (1993) Osservazioni sulla distribuzione della teriofauna nella pianura veneta centro-orientale. *Boll. Mus. civ. St. Nat. Venezia*, **42**, 165–193.

Bon, M., Cherubini, G. and Stival, E. (in press) Il progetto atlante degli uccelli nidificanti in Provincia di Venezia: dati preliminari. Presented at the *2° Convegno dei Faunisti Veneti*", Padua, 25–26 October 1997.

Bon, M., Scattolin, M., Anoe, N., Borziello, G. and Busso, C. (1996) *Il Forte di Carpenedo - Flora fauna e ambiente naturale*. Venice: Arsenale Editrice.

Bonometto, L. (1995) *Un ambiente naturale unico. Le spiagge e le dune della penisola del Cavallino*, Venice Municipality.

Borgoni, N., Richard, J. and Semenzato, M. (1994) Gli Anfibi e i Rettili delle cave dismesse dell'entroterra veneziano. In: *Proceedings of the 1° Convegno dei Faunisti Veneti*, pp. 39–41, Montebelluna.

Brichetti, P. and Gariboldi, A. (1992) Un Valore per le specie ornitiche nidificanti in Italia. *Riv. Ital. Ornithol.*, **62**, 73–87.

Calzavara, D. (1979) Le casse di colmata della laguna media, a sud di Venezia – II. Note preliminari sulla vegetazione della Cassa D-E. *Lavori Soc. Ven. Sci. Nat*., **4**, 81–88.

Canestrelli, P. (1979) Le casse di colmata della laguna media, a sud di Venezia – IV. La fauna ortotteroidea della Cassa D-E. *Lavori Soc. Ven. Sci. Nat.*, **4**, 89–91.

Caniglia, G. (1978) Tracce di vegetazione spontanea in un settore litorale del Cavallino (VE). *Boll. Mus. civ. St. Nat. Venezia*, **29**, suppl., 169–192.

Caniglia, G. (1981) Il bosco di Carpenedo. Lav. Soc. Ven. Sci. Nat., 6, 151-158.

Caniglia, G. and Salviato, L. (1983) Aspetti vegetazionali sulla colonizzazione di un ambiente di bonifica della laguna di Venezia. La cassa di colmata B. *Atti Mus. Civ. St. Nat. Trieste*, **35**, 91–120.

Caniglia, G., Borella, S., Curiel, D., *et al.* (1992a) Distribuzione delle fanerogame marine (*Zostera marina, Zostera noltii, Cymodocea nodosa*) in laguna di Venezia. *Lav. Soc. Ven. Sci. Nat*., **17**, 137–150.

Caniglia, G., Mondin, F. and Carpene, B. (1992b) Aspetti foristico-vegetazionali di un lobo di meandro del parco del Sile (S. Michele Vecchio – Venezia). *Lav. Soc. Ven. Sci. Nat.*, **17**, 137–150.

Carrada, G. (1990) Le lagune costiere. *Le Scienze*, **264**, 32–39.

Cesari, P. and Orlandini, M. (1984) Il Bosco di Carpenedo (Venezia) – 2. Notule corologiche e sistematiche sulla malacofauna di un aobiente relitto dell'entroterra veneziano. *Lav. Soc. Ven. Sci. Nat.*, **9**, 131–176.

Cherubini, G. and Panzarin, F. (1994) Il Fratino Charadrius alexandrinus nidificante lungo i litorali della provincia di Venezia. In: *Proceedings of the I Convegno Faunisti Veneti*, pp. 111–112, Montebelluna.

Favero, V., Parolini, R. and Scattolin, M. (1988) *Morfologia storica della laguna di Venezia*. Venice: Arsenale Editrice.

Gehu, J.M., Scoppola, A., Caniglia, G., Marchiori, S. and Gehu-Franck, J. (1984) Les systèmes vegetaux de la Côte Nord-Adriatique italienne, leur originalité a l'échelle Europèenne. *Documents Phyvosociologiques N.S.*, **8**, 485–558.

Giordani Soika, A. (1949) Studi sulle Olocenosi VII. Notizie e considerazioni preliminari sulla fauna sottobasale delle praterie di Zostera della laguna di Venezia. *Proc. Soc. Nat. Matemat. Modenesi*, **80**, 1–15.

Giordani Soika, A. and Perin, G. (1974) L'inquinamento della laguna di Venezia: studio delle modificazioni chimiche e del popolamento sottobasale dei sedimenti lagunari negli ultimi vent' anni. *Boll. Mus. civ. St. Nat. Venezia* **26**, 25–68.

Guelorget, O. and Perthuisot, J.P. (1983) Le domaine paralique. Expressions géologiques, biologiques et économiques du confinement. *Traveaux du laboratoire de géologie*, **16.10**

Hieke-Merlin, O., Menegazzo Vitturi, L. and Semenzato, G. (1979) Contributo alla conoscenza dei sedimenti superficiali della laguna veneta. *Atti. Ist. Ven. Sci. lett. Arti.*, **137**, 35–51.

Mainardi, D. (1996). L'impatto dell'uomo sulla natura. *Univ. Ca'Foscari di Venezia, Prolusioni,* pp. 35.

Pignatti, S. (1952-53) Introduzione allo studio fitosociologico della pianura veneta orientale con particolare riguardo alla vegetazione litoranea. *Arch. Bot.*, **28**, 265–329; **29**, 1–25, 65–98, 129–174.

Pignatti, S. (1959) Il popolamento vegetale. *Boll. Mus. civ. St. Nat. Venezia*, **12**, 61–142.

Pignatti, S. (1966) La vegetazione alofila della laguna veneta. *Mem. Ist. Ven. Sc. Lett. Arti, cl. Sci. Mat. Nat.*, **3**, 1–174.

Raffone, G. (1991) Ricerche ditteriologiche nelle cave di Gaggio di Marcon (Venezia). VI. Fam. Empididae, Hybotidae, Asteiidae, Syrphidae, Opomyzidae, Anthomyiidae, (Gen. Lispe), Calliphoridae (Gen. Lucilia). *Lavori Soc. Ven. Sci. Nat.*, **17**, 3–12.

Rallo, G. (1978) Le casse di colmata della laguna media, a sud di Venezia. (Nota preliminare sull'avifauna). *Lavori Soc. Ven. Sci. Nat.*, **3**, 55–66.

Ratti, E. (1979) Le casse di colmata della laguna media, a sud di Venezia – V. La coleotterofauna della Cassa D-E. *Lavori Soc. Ven. Sci. Nat.*, **4**, 115–169.

Ratti, E. (1984) Il Bosco di Carpenedo (Venezia) – 3. Osservazioni sulla coleotterofauna di un lembo relitto di foresta planiziale. *Lav. Soc. Ven. Sci. Nat.*, **9**, 187–191.

Ratti, E., De Martin, P. and Zanella, L. (1995) I Coleotteri Carabidi di un lobo di meandro del fiume Sile presso Quarto d'Altino (Venezia). *Boll. Mus. civ. St. Nat. Venezia*, **44**, 63–78.

Richard, J. and Semenzato, M. (1988) Il bosco di Carpenedo (Venezia) – 4. Osservazioni sugli anfibi e rettili di un lembo relitto di foresta planiziale. *Lav. Soc. Ven. Sci. Nat.*, **13**, 103–114.

Roccaforte, P., Sirna, G. and Bon, M. (1994) Il Bosco di Carpenedo (Venezia) - 6. Osservazioni sull'avifauna di un lembo relitto di foresta planiziale. *Boll. Mus. civ. St. Nat. Venezia*, **43**, 221–230.

Rose, P.M. and Scott, D.A. (1994) *Waterfowl Population Estimates*, International Waterfowl and Wetlands Research Bureau Publication, 29, Slimbridge, 102 pp.

Scarton, F. and Valle, R. (in press) L'avifauna nidificante sulle barene della laguna di Venezia: anni 1989–1995. In: XIII Convegno del Gruppo di Ecologia di Base G. Gadio: Aspetti ecologici e naturalistici dei sistemi lagunari e costieri, Venice, 25–27 May 1996.

Scarton, F. and Valle, R. (1997) Laridae e Sternidae nidificanti in laguna di Venezia: anni 1995–1996. *Riv. Ital. Ornithol.*

Scarton, F., Borella, S. and Valle, R. (1996) Prima nidificazione di Beccapesci *Sterna sandvicensis* in laguna di Venezia. *Riv. Ital. Ornithol.*, **66**, 87–88.

Sfriso, A., Marcomini, A., Pavoni, B. and Orio, A.A. (1990). La Laguna di Venezia come caso esemplare. "*Inquinamento*", **4**, 62–79.

Stival, E. (1996) *Atlante degli uccelli svernanti in provincia di Venezia. Inverni dal 1988/89 al 1993/94*, Centro Ornitologico Veneto Orientale, 213 pp., Montebelluna (TV).

Valle, R., Scarton, F., Borella, S. and Tiloca, G. (1994) Nidificazioni di Avocetta (*Recurvirostra avosetta*) nella laguna di Venezia. *Lavori Soc. Ven. Sc. Nat.*, **19**, 99–102.

Valle, R., Scarton, F., Tinarelli, R., Grussu, M., Utmar, P. and Borella, S. (1995a) Primo censimento della popolazione di Pettegola (*Tringa totanus*) nidificante in Italia. *Suppl. Ric. Biol. Selvaggina*, **22**, 601–605.

Valle, R., Piacentini, D., Scarton, F., *et al.* (1995b) Status e distribuzione della Volpoca *Tadorna tadorna* nidificante in Italia. *Avocetta,* **19**, 163.

Var. (1985) *Un parco nella Laguna di Venezia*. Venice: Arsenale Editrice.

Var. (1986) *laguna. Conservazione di un Ecosistema*. Venice: Arsenale Editrice.

Var. (1991) *Le alghe della laguna di Venezia, Vol.I*. Venice: Arsenale Editrice.

Var. (1992) *La laguna – Ambiente Fauna e Flora*. Venice: Corbo and Fiori Editrice.

Vatova, A. (1949) Caratteri di alcune facies bentoniche della laguna Veneta. *Nova Thalassia*, **1**, 1–13.

Vatova, A. (1981) Recherches comparatives sur les "valli" salées de peche de la Haute Adriatique, *Rapp. Comm. int. Mer Médit.*, **27**, 93–94.

CHAPTER 3

On the Natural Equilibrium of the Venice Lagoon (Will Venice Survive?)

Andrea Rinaldo

1. Introduction

You can call it a gamble or, as André Chastel once put it, '...the Venetian challenge: the central episode of the crisis of modern civilization': is there a way to preserve Venice, as a living treasure, for future generations? Difficult choices lie ahead, because the combined effects of sea level rise associated with global warming and of local and non-local subsidence will increase the frequency and height of tidal flooding to the extent that Venice will probably become uninhabitable within the next 100 years unless structural measures are taken. If a conflict between the physical and social welfare of the city and the strict conservation of its current environment arises will we be able or willing to choose?

The Venice issue is indeed becoming a case study of paramount importance, not only for the intrinsic importance of the historic, artistic and architectural value of the city which *per se* commands global attention, but also because it is a paradigm of a new complexity in the interactions among economics, society and the environment. It is such complexity that motivates the need for defining and pursuing environmentally sustainable development, i.e. economic growth maintaining adequate quality of life and of the environment. Its definition, however stretchable, must imply a reference, natural state to be maintained or to be aimed at. But what if no 'natural' state of the environment exists? What about ecosystems, like the Venetian lagoon, that have been radically modified through the centuries to serve the needs of social and economic developments? A new definition of environmental sustainability is thus needed.

The above questions are, in my view, at the heart of key choices that humankind will face in the near future – should we pursue social and economic

I. Musu (ed.), Sustainable Venice, 61–93.

prosperity, the strict physical preservation of the current configuration of a complex ecosystem, possibly a byproduct of complex coevolutionary processes involving both genes and cultures, or should we allow unleashed natural evolution and risk the loss of human artifacts, possibly unique? In some yet wider sense, does the notion of natural equilibrium, in the static maintenance sense it suggests (implied more or less explicitly by the 'do nothing' options ventilated by conservationists), make any sense in natural evolutionary phenomena like those at work in molding the venetian environment?

My main conclusion, facing the relevant history and science, is that Venice and its environment are essentially an artificial system that will be preserved only after major interventions. I also favour the view that Venice must survive from the waters in the waters, irrespective of (and without much regard for) 'natural' evolution, whatever intertwined shapes should we design for the city, its mailand territories and ecosystem.

2. Of science and politics

Sustainable development programmes propose to reach the objective of a collective undertaking of responsibility for important environmental issues through the activation of educational and participatory mechanisms. There is a wide-felt need for such an operation of technically sound[1] sensitisation, particularly regarding the problem of the Venice environmental system (the city, the lagoon and its contributing mainland areas). It could be thought – wrongly, I believe – that everything has already been said or written about this problem. Indeed, over only the last 15 years, the technical resources which have been made available in order to provide critical analysis with suitable instruments have substantially changed the reference framework concerning the drainage basin–lagoon–sea system. This has made a total contribution to the body of knowledge formed over the centuries which has been quantified at about 80%.[2] Technical choices and historical errors[3] were based, at least in the past, on a type of engineering founded on observation rather than on reason, on heated debates, on many corrections, even fundamental ones, to the measures adopted from time to time in order to meet the changed socio-economic conditions. As a result, a critical comparison of ideas and proposals, although limited, can perhaps set out to draw attention to some preconceptions and provide a suitable basis for the debate.

The process of constructing possible consensus between social groups on a plan for urban and environmental sustainability must take some preliminary definitions of the nature of the relationship between city and its environment into consideration. The sustainability of development implies the attainment and maintenance of a model of human life which is organised in harmony with its environment. Instead, the history of Venice teaches us the opposite; here, there is the resolute definition of an environment which moulds itself to the city's needs for living and development, with nothing or nearly nothing left to natural evolution. Therefore, it is not a matter of a traditional relationship between development and environment.[4] To consider the fundamental issue of physical protection from a historical perspective would, perhaps, mean radically setting the problem of nature conservation, specifically that of the lagoon, against the conservation of a resource. The very title of the chapter assigned to me reflects a conviction, almost a hope. This is that, within the dynamics of a complex system which is 'open' to the injection of energy and mass from the outside, the general notion of equilibrium can prevail. This notion is understood as an image of a static nature which, forced by different agents, thus halts its evolution in a desirable state and, in principle, can, therefore, be conserved. It is no coincidence that this idea motivates environmentalists and conservationists since the natural world usually changes very little in a human lifetime and, as a result, the concept of equilibrium misleadingly seems extremely natural and intuitive.[5]

If nature is in equilibrium, how have we ended up in it? Does a preferential state of nature perhaps exist, such as the way in which we now see it from the evolutionist perspective? It is commonly accepted today that the natural world is permanently far removed from equilibrium and is self-organised into possible states towards which the dynamics of the system is attracted by the invisible hand[6] driving evolution.[7]

Moreover, it is obvious but useful to remember that any radicalisation in the search for consensus between social groups appears to be the least advisable approach to adopt. In a particularly evolved context, such as that of the lagoon ecosystem, and with an informed public which has already acquired a large amount of the basic information on the city's problems and those of its lagoon, any attempt at forced imposition appears damaging and useless.[8]

This chapter is devoted to making a (limited and partial) critical account of the modifications to the lagoon morphology[9] which have taken place, and of their implications for, and limitations to, any future economic specialisation of the city. Furthermore, it examines the way in which the physical protection of the city can have repercussions on its sustainable development. A necessary task in examining the state of the art of the proposals and knowledge is that of not siding with any one particular choice, but instead, of critically examining its consequences. Thus, no proposal must be demonised, especially by looking back, instead of forward, to the real or presumed milieu which brings it about; the presumed or real interests which it defends; the model of development which it underlies or sustains. Every proposal must be examined with suitable tools, without claiming the superiority of one discipline over another: each to their own. As is obvious, the decision lies with the politician and the population he is answerable to. It is up to us to render this decision transparent and reasonable, meditated and unhurried and not mysterious or misleading. The essential instrument remains that of critical doubt along with rigorous historical, technical and economic analysis. Research on the ways of conserving and protecting Venice and its environment has to proceed by establishing common ground for all parties involved in the process, without discriminations of any kind: frankly this seems unthinkable at present.

3. Natural equilibrium?

The concept of natural equilibrium, in the static sense of maintenance which the notion suggests, does not feature at all in any natural evolutionary phenomena.[10] However, if such a statement is true in general it is even more evident in the case of the Venice environment (its lagoon and drainage basin), which has been the object of much intervention for centuries. It has been maintained completely artificially only at the price of decisive transformations carried out by man, somewhat scientifically in some cases, by trial and error in many others.

The following briefly summarises the main human intervention and its essential chronological order:[11]

(i) the large diversions of rivers (Piave, Sile and Brenta). These constitute a set of interventions begun in 1324, conceptually started up over 30 decisive years in the 1500s and concluded in 1860 with the definitive diversion of the Brenta into the Brondolo lagoon which was then rapidly filled in by silt.[12] The Piave bed was diverted (1534, 1565 and 1642–1664), thereby creating the Taglio di Re from San Donà to Eraclea, and its flow was directed into the sea to its current site at Porto di Cortellazzo. Together with the Piave, the Sile was also diverted to the sea (1683), in keeping with its current course along the boundaries of the North lagoon. However, the colossal operation of the Taglio of Porto Viro of the deltaic branch called Po delle Fornaci must not be overlooked. This was fought for at the beginning of the 1600s by Luigi Groto and is the cause of the silting up of the lagoon openings with sand as a result of the marine currents. These works permitted the lagoon basin to be maintained intact as a transition region because, by isolating Venice, it granted the city more secure protection and defence. On the other hand, they eliminated the vast majority of the solid material loads which originally modelled the lagoon (all the remaining courses of freshwater – from North to South, the Zero, Dese and Marzenego, which was also subjected to a partial diversion in 1507, apart from a relict of the Brenta called the Naviglio, of little hydrological significance – are of subsurface origin and do not carry large quantities of sediment even under extreme hydrological conditions), thus reversing the trend towards filling in and maintaining the lagoon clean and crossed by lively currents but also condemning the lagoon basin to its transformation into a sea reach;

(ii) the sea defence works, or rather the seventeenth century reinforcement of the coastal strips (*murazzi*, sea walls), which were designed to protect the fine sandy strips from the overflow of stormy seas, the very phenomenon which originally brought about the formation of the lagoon;

(iii) the changes in the lagoon boundaries: the closure of inlets of various dimensions and locations; the planning, maintenance and subsequent destruction of numerous lagoon salt pans; the embankment of the fish-farming valleys; the planning and creation of small bays, reclamation works of various natures and functions, banks and morphological inteventions (*traversagni, drizzagni, tagliate, scomenzere* – meaning cuts, rectification of meandering tidal channels, fillings, reclamations etc.); and, naturally, the excavation *ex novo* of numerous navigable channels (I am not referring to this century nor to the notorious *Canale dei Petroli*, the hated oil-tanker channel that was dredged to divert the oil traffic from the waterways close to the city) from the Drizzagno Channel to the Cunetta Channel, from the Santo Spirito Channel to the Nuovo Channel. The complex artificial navigable system of communication between the basins of Lido and Malamocco, the Orfano Channel – Santo Spirito Channel system, appears to be of particular importance at the beginning of the 1700s;

(iv) the continuous maintenance works on the sandbars at the mouths which limited the 'Serenissima' (literally, the 'most serene') Republic's maritime traffic and the maintenance of the otherwise non-navigable lagoon depths. For example, in Cristoforo Sabbadino's time, the depth at St. Nicolò was 4.20 m (Sabbadino, 1930). In 1723, with a reduction in maintenance due to the Serenissima Republic's contracted economic and military power, the opening at St. Nicolò could be entered at a depth of little less than 4 m (Zendrini, 1723). At the beginning of the 1800s the minimum useful depth was a little less than 3 m (Denaix, 1811) due to the progressive filling in of the lagoon;

(v) the construction of the breakwaters, the jetties and and the dykes which extend the Malamocco inlet into the sea (1845–1865); the demolition of the ancient inlet structures at Lido, St. Erasmo and Treporti, which were then joined in the current structure of the Lido inlet (1845–1865); the twentieth century structuring of the Chioggia inlet, begun in 1914 and concluded in the 1950s. These works caused the elimination of the sandbanks and bars at the inlets, thereby guaranteeing adequate and stable depths for navigation at the price of an increased volume exchanged between sea and lagoon. The corollary of the enlivening of the tidal currents is the wearing away of the lagoon shores and margins (the partial collapse of the St. Andrea Fort is a spectacular case, but the very structure of the city's crumbling foundations cannot have avoided feeling the effects of this) as well as the demolition of the sandbanks and of the familiar lagoonal topography.

Many of the works which have taken place over the centuries have had serious negative side effects and have had major environmental impacts, even at the distance of centuries. The early hydraulic engineers cannot be blamed for not having considered the secondary and crucial effects of many works carried out since they had only intuition to support them and *a posteriori* empirical proof of these works' effects. It would be more difficult, however, to exonerate the modern ones because of the changed state of knowledge, of capacity for prediction and of the instruments for intervention. The crucial and fateful decisions for the Venetian environment, however, have distant origins. In fact, the natural evolution of a complex system such as the lagoon is designed by the prevalence of the activity of fluvial contribution or erosion due to wave action or that of the tidal currents caused by the sea or by their balancing. All of this is sustained by eustasy and subsidence, The most important physical factor in conditioning its evolution (Gatto and Carbognin, 1981) is the definitive exclusion of the rivers concluded in 1896.[13] Therefore, the lagoon's evolution appears to be strongly driven towards the *facies* of a sea reach. This is a process which has been considerably accelerated by, though not caused by, the construction of the large navigable channels (1925, Vittorio Emaneule Channel; 1968 – Malamocco-Marghera Channel (*Canale dei Petroli*)).

Nevertheless, it is evident that the concept of natural equilibrium, for a system which requires such major interventions, is absurd. Instead, we must be recognise that, throughout the centuries, Venice's rulers have had to continually intervene and tamper with the lagoon in order to fashion it into a liveable and vital environment in accordance with the demands of the city's economic and social development (Dorigo, 1972).

Venice and and its Lagoon: an essentially artificial whole. Furthermore, if Venice is to be saved from the water, it must, without doubt, be saved in the water (Bettin, 1993), whatever order the city and its territory possess today and whatever order should be designed for their future.

4. Of the lagoon equilibrium (and of common prejudices)

The hydrodynamic problem of tidal propagation is often discussed in terms influenced by an interpretation of the famous sixteenth century aphorism, of unknown origin (Adami and Baschieri, 1995), '*gran laguna fa gran porto*' (large lagoon makes for large inlet). According to this, the exchange of volumes and the flow of water between sea and lagoon is solely determined by the relationship between the surface area of the aqueous part of the lagoon and the size of the inlets. The implications of the aphorism regard both the hydrodynamics and the morphology of the lagoon basin, and provide an image of the lagoon as one which is filled up and emptied like a vase, communicating with the sea through its inlets, and dependent upon the fluctuations in the sea level of the Upper Adriatic. Since what surrounds the inlet sections cannot be considered as fixed, but is rather mobile under the influence of current dynamics, it is reasonable to imagine that the section of the inlets adapts to the flow being conveyed (if the velocity of the water is too high then the inlet will tend to be eroded until it reaches a size which is large enough to generate velocities which are compatible with the stability of the bed. In the opposite situation, the sediments will tend to be deposited, reducing the section of the inlet). This simple and understandable hypothesis on the functioning of the lagoon, known as the static hypothesis, is completely compatible with the sixteenth century saying. This supports the interpretation given to the correlation between the reclamation of lagoon areas, the excavation of navigable channels and high tides which are generated by the larger volumes exchanged between the sea and lagoon which would then produce greater fluctuations in level.

The static hypothesis, however, is misleadingly inaccurate.[14] As has been noted, the lagoon water level does not rise and lower nor maintain itself horizontal as if in a bowl, but propagates wave-like from the inlets to the boundaries (*la conterminazione* or *il contermine*) with delays which may be some hours. The nature of a tidal wave's propagation is influenced by many factors which define the relationship between the celerity of tidal propagation,[15] the inertia of the mass involved and flow resistance phenomena (Ippen, 1966). In propagating through an inlet along the lagoon channels (as far as the boundaries and interacting in the areas of the hydraulic watersheds), the tide wave is deformed, either damped or magnified, according to a possibly resonant relationship between local inertia and resistance and the characteristics of the wave at sea. (Dronkers, 1964).The validity of the relationship between the surface area of tide expansion and the inlet dimensions is not general. If a lagoon is very large, it could be that the inlets tend to enlarge instead of narrow when the extent of the lagoon basin is reduced.[16] Other kinematic and dynamic hypotheses are applicable. From these it is possible to deduce the particular conditions in which a lagoon operates from a hydrodynamic perspective. In brief, a large lagoon means a large inlet, only for lagoons which are sufficiently short and in which the propagation time between sea and lagoon boundaries does not exceed around a quarter of the forcing tidal period.

Today, the propagation of tides in arbitrarily complex lagoon basins is studied with accurate and mathematical numerical models. Hydrodynamic mathematical models define from first principles the dynamic balance between mass and energy which includes a complete description of the contributions of momentum, local inertia, gradients of free surface, flow resistance at the bottom and wind force. The boundary conditions of the mathematical model are the instantaneous sea level and possibly the freshwater inflows from the main land.[17,18] These models are able to describe the field of motion's complex geometry[19] in great detail (unimaginable at the first definition of the lagoon's behaviour with the static method) (Ghetti *et al.*, 1970). Today, both a complete calibration of the parameters of the models, conducted on the basis of the tidal events registered at different points throughout the lagoon[20] and of measurements of currents recorded in the field (Istituto di Idraulica dell'Università di Padova, 1979; Consorzio Venezia Nuova, 1989), and a detailed comparison between different approaches and models allows a precise assessment of the effects of structural intervention on the tidal propagation and, therefore, of the cause and effect relationship in terms of past interventions (Consorzio Venezia Nuova 1989).

On the basis of such a mass of results, it could be thought that there must be a general consensus held on the cause and effect relationships of the various works on the lagoon hydraulics which have taken place throughout the centuries. It does not appear to be so.[21] The role played by this century's large lagoon channel excavations in acting as a decisive cause of hydrological and morphological damage is particularly debated.[22] To contribute to clarifying different interpretations is, perhaps, a useful exercise, at least with regard to some lagoon hydraulics issues which are not immediately obvious.

It is difficult to maintain that the increase in the frequency of high water observed since the 1960s is mainly due to tampering with the lagoon, to the careless works carried out during the period after the Second World War and, *in primis*, to the large navigable channels.[23] It is certainly true that in a tidal channel with a particular form and length, tides can be produced which are amplified with respect to forcing oscillation at the inlet, due to the phenomenon known as resonance, as a result of the combined effect of the wave inertia, resistance to the motion and the characteristics of the forcing oscillation (Dronkers, 1964). There can thus be distinguished the 'short' lagoon (far from conditions of resonance for any value of the flow resistance and of the forcing tides) and the 'long' lagoon (for which the propagation times are long in relation to the tidal period. These are strongly dependent on the resistance to motion which conditions the deformation of the tide wave during its propagation); lagoons which are close to the condition of resonance which, in linear dynamic models, are characterised by a propagation time between sea and lagoon edge to the order of one-quarter of the tidal oscillation period. In resonant conditions, the amplification of the wave heavily depends on the resistance to overall motion. The Venice lagoon is not far from resonant conditions for significant tides,[24] since propagation times of little more than 2 h have been measured and calculated for the Chioggia and Malamocco basins and of little less than 3 h for the north-eastern end of the Lido basin. Moreover, the excavation of the navigable channels has brought the propagation conditions closer to resonant conditions. Therefore, it is credible that the considerable modifications in tidal propagation in the lagoon made by the excavation of the navigable channels have caused, for certain tides, an exacerbation of the phenomenon of high water. However, in order for this exacerbation to be significant, the length of the tide's course has to be comparable to the wavelength of the sea's forcing oscillation (Dronkers, 1964). For example, the wavelength of the semidiurnal tide is to the order of 150 km. In relation to this, any modification inside the lagoon is unable to take away its hydraulically 'short' character which prevents significant amplification.[25] The evaluation of the extent of the amplification is a matter for debate. Studies and statistical calculations of interest (Tomasin, 1974; Pirazzoli, 1982, 1987) carefully consider the empirical values of the height of high water in Venice over the last century. Working with moving averages on intervals of diverse lengths, in order to filter out astronomical elements, the increase in high water in this century is estimated at about 40 cm, 27 cm of which are attributed to the relative rise in average sea-level due to subsidence and eustasy[26] and 14 cm to hydrodynamic factors, about 10 cm of which is the average contribution of tidal amplification caused by the navigable channels. Furthermore, a comparison between the data registered at the stations at the South Lagoon Dam (in the sea) and the Punta della Salute shows a rise in annual maximum tides inside the lagoon of about 6 cm. In the author's opinion this is not attributable to differential subsidence.

The different hydrodynamic mathematical models precisely define the influence of the navigable channels on the levels and, at the same time, permit the evaluation of the little-considered effects of intervention carried out at the inlets. The models use as a reference, for example, the hydrodynamic behaviour of the above-mentioned Denaix lagoon, which is evaluated by propagating the tides on the nineteenth century bathymetric scale. Today, the tidal peaks immediately inside the inlets are slightly reduced with respect to the sea because of the amplitude of the tidal openings, and undergo a slight amplification or a further reduction in the subsequent propagation, depending on the type of tide (i.e. whether inertial or dissipative effects prevail) (D'Alpaos, 1992). It is interesting that the effects of the (nineteenth century) structuring of the inlets are greater in the long-term than those produced by the navigable channels in this century. Nonetheless, the statistical estimation of 10 cm net increase owing to the dredging of the navigable channels appears excessive.[27] Instead, the nineteenth century lagoon, dominated by resistance and characterised by reduced exchange with the sea, significantly reduced, on average, the tidal peaks.[28]

Whether the amplification of the high tide is 10 or 2–3 cm actually changes the situation very little. Personally, I hold the lower estimate to be more probable (despite the certitude of Pirazzoli, 1987) as a result of certain deficiencies in the statistical analysis. Indeed, statistics cannot take account of the fact that the subsidence of the whole lagoon basin by more than 20 cm reduces, and not only slightly, the hydraulic roughness of the lagoon, particularly in its shallows. The contribution of the diminished resistance to motion is estimated as 1–2 cm depending on the variations in the tide's gradient (D'Alpaos, 1992). Furthermore, differential subsidence cannot be ruled out for the part related to the artificial extraction from subsoil.[29] In addition, the role of statistical manipulation in terms of the important and atypical meteorological elements, of the differing morphological response over time, of the errors in measurement which cannot be uniform over a period of 100 years and of yet other details is not completely clear.[30]

The absence of a founded correlation between reclamation, large channels and high tides with regard to exceptional events is clear.[31] Indeed, if the tide in the Adriatic remains over 1.50 m above average sea-level for more than 15 h (as in the extreme event of 4th November 1966), a virtually identical elevation will be reproduced within the lagoon independently of the nature of the propagation (or rather, of the depth and width of the lagoon channels, even if the lagoon were extended as far as Padua, thereby showing no correlation between lagoon water body and high tide). In the absence of artificial forms of control, the exceptional high tides have the city at their mercy.[32]

A current opinion[33] sustains that, above all, the lagoon, as bequeathed to us from the 1800s, must be fortified and conserved and that structural protection measures must be pursued by essentially abandoning the control of exceptional high waters and instead, abating the average-high tides with wide-area morphological measures and with the opening of the fish-farming valleys to the expansion of the tides.[34] The wide-area measures consist of the reduction in the depths at the inlets; the restoration of the system of filtration and resistance to tidal propagation of salt marshes, shoals and bumps (dossi); the closure of the straight reach of the *Canale dei Petroli* (the Malamocco-Marghera Channel). As far as the inlets are concerned, they should be directed differently[35] and reduced to modest depths comparable to those of the eighteenth century lagoon in order to re-design Venice's role as a port in a different direction. The stated aim is that of reversing the degenerative processes in progress by reducing the velocity of the tidal currents and their erosive powers as well as reducing the tidal peaks in the historical centre as a result of the reduced volume of sea water entering and the increased resistance to the propagation of the tide.

I do not deny the respectability of these proposals. Nevertheless, their technical feasibility and their effects on the physical system[36] are predictable[37] and must be clearly pointed out.

Extensive morphological measures (particularly reduction in depth and in hydraulic sections as well as the local increase in roughness) are of environmental interest if they follow morphologies which are close to natural ones. However, they are technically costly[38] and certainly insignificant[39] for the reduction of exceptional tidal peaks. For the average-high tides, the undeniable benefit depends on the how steep the tidal oscillations are.[40] In addition to the fact that a reversal in the trend of morphological degradation cannot be considered guaranteed,[41] the variations in channel depth and in the depth of inlet sections increase sedimentation and the possible cost of dredging and maintaining the (potentially) navigable channels. Furthermore, water quality problems would undoubtedly increase as a result of the desired reduction in exchange between sea and lagoon.[42] Therefore, the vast restoration of the lagoon ecosystem which would be produced by the wide-area morphological measures is fascinating but problematic from a technical point of view. On the one hand, this is because it does not eliminate, even if considerable expense is involved, any of the really dangerous effects on the city. On the other hand, it is because it does not substantially modify the degenerative processes taking place and it becomes a kind of 'torment of Tantalus'; destined to perpetual reconstruction and maintenance work in order to endure.

The opening of the fish-farming valleys alone to the expansion of the incoming tide does not create any significant hydraulic benefit for the average-high tides nor cause any reduction in the high tides in exceptional events (see Istituto di Idraulica G. Poleni dell'Università di Padova, 1970; D'Alpaos and

Di Silvio, 1975; Adami, 1992; D'Alpaos, 1992). Therefore, only for other reasons[43] could it be suggested to use them differently since their hydrological public utility should be considered non-existent. Significantly, the fish-farming valleys play a completely marginal role with respect to the current's regime and to the levels in the historical centre, as they do even for the nineteenth century Denaix lagoon which is characterised by a notably greater total surface area than the lagoon basins (D'Alpaos, 1992). It is interesting to observe that the increase in tidal propagation times following an opening of the fish-farming valleys would bring about a slight reduction in the volume exchanged through the Lido inlet and a slight increase in that exchanged through the other two inlets (Di Silvio, 1992). It should also be observed that certain descriptions[44] of the effects of closing the valleys to tidal expansion are impressive but technically ludicrous and are of no use to otherwise largely supported theories relevant to the future of the city's development.

The control of the exceptional high tides cannot be achieved without controlling the inlets with fixed or mobile installations. All models and evaluations (national and international) agree on this point.[45] The tenet of the design of control works is based on the idea of temporarily isolating the lagoon from the sea during periods of high water. Many different solutions, with configurations of mobile installations, fixed installations or a combination of one and the other, have been studied (Consorzio Venezia Nuova, 1989). All these pose different technical, economic and management problems. The concessionaire's final choice of project, which was produced with the intent of making subsequent refinement, provides for mobile barriers which can disappear completely. In recent months the project has been subjected to an environmental impact assessment procedure which is an examination of great importance. This is not the place for a discussion on the strengths and weaknesses of the colossal project for the mobile barriers at the inlets. However, it is obvious that the engineering which has been mobilised by the state concessionaire, the Consorzio Venezia Nuova, cannot be ignored (as seems to happen, particularly in the city), or underestimated. Nor can I say that I agree with those who maintain that these installations at the inlets are specially to the structure and depth of the present lagoon channels as well as to an outdated and dangerous port and industrial model (Rosa Salva, 1997). In fact, the mobile barriers' use and function can be justified in relation to any lagoon morphology.[46] Therefore, from this point onwards the problem cannot be considered a technical one, but falls within the domain of the politics of strategic decisions concerning the city. Let me restate the main points once more: if the control of flooding from exceptional tides is an important problem for the city, its only possible solution is the temporary isolation of the lagoon from the sea by means of intervention at the inlets.

A further, and complex, problem is the regulation of future, possible mobile barriers. Statistical regularisation permits the evaluation of the return time of tidal levels which exceed a predetermined height. In the present altimetrical conditions, for heights greater than 1.00 m a.s.l. (at which 5% of the city is flooded), the average frequency is about seven events a year (with average duration estimated to be from 16 to 46 h); the maximum observed (in 1979) is about 60 times (Ghetti, 1984a, b). The return time for heights greater than 1.40 m a.s.l. is about 6 years, and greater than 1.60 m, about 20 years. Heights greater than 0.80 m occur about 40 times a year (Ghetti, 1976). These values, relative to the present average sea-level, permit the evaluation of some of the complex issues which govern the possible regulation of the temporary isolation of the Lagoon from the sea with caveats, see next paragraph.

5. Eustacy and subsidence

The primary causes of the increase in the frequency of high tides in the historical centre, which are also a lethal danger to the city's fate, are the concurrent phenomena of subsidence and the increase in average sea-level (eustacy). Today these phenomena are only connected to natural processes but in the past they were exacerbated by the improvident works of man, although not carried out in Venice or in its lagoon.[47] It is easy to grasp the correlation between the reduction in average elevation of the city and the increase in high water events. The total reduction in this century is reliably estimated at about 23 cm which corresponds to a rate of eustacy of about 1.1 mm/year which, coupled with subsidence, has caused the total velocity of Venice's progressive yielding to be about 2.5 mm/year[48]. A single comforting fact must be highlighted: when a serious campaign of scientific information indicated the certain theoretical and empirical correlation between subsidence and the extraction of water from the subsoils of the lagoon edges and of the drainage basin (see Gatto and Carbognin, 1981), the industrial wells, which were the main culprits of such extraction, were rapidly closed and substituted with an industrial pipeline from the Sile to Porto Marghera. The practical result was the elimination of the so-called anthropic element of the subsidence caused by extraction and subsidence then settled once again at values characteristic of geological compaction (3–7 cm/year). I do not relinquish hope that this type of consensus and effectiveness can be repeated in relation to the hydrological and morphological issues.

Recent work at six archaeological sites in Venice suggests yielding trends in relative sea levels from A.D. 400 through 1990 (Ammerman *et al.*, 1999), complementing established archaeological estimates (Dorigo, 1983). Through such work, the long-term rise in relative sea level and predictive scenarios are derived (see, for a critical comparison, Co.Ri.La, 1999). I notice, however, that the ensuing scientific debate has been misled and confused by contributions (e.g. Ammerman and McClennen, 2000) tainted by political agendas.

However, the forecasts for eustacy in the third millennium are bleak.[49] In fact, the average estimate from the general circulation models, the state of the art of knowledge in matters of global change (climate change on a world-wide scale), and from limited area models, predict an increase from 20 to 50 cm for the year 2100.[50] Venice would be lost in the passing of two or three generations. This will prevent it from being saved in the water because if the choice of no intervention prevails (the so-called zero option) in 50 years it will be too late for non-radical intervention and the only course to be taken will be to permanently isolate the city from the sea.[51] Therefore, as stated by Chastel (1990), the Venetian challenge, the central episode of the crisis of the modern world, will be lost.

6. Lagoon morphology

Another crucial issue, the relationship between lagoon morphology and human intervention, has to be examined carefully, both because of its implications for the physical 'equilibrium' of the Venetian environment as well as for an examination of the evolutionary trends, in progress or which can be predicted, with changed policies.

The natural environment which we call the lagoon is the product of the concerted action of many physical agents which determine the movement of sediments by fluvial, lagoon (connected to the tidal currents) and marine dynamics. That is, the loads of solid material from water courses and from the marine currents: wave action and the coastal marine currents; wave action and the tidal currents in the lagoon; the phenomena of bradyseism (subsidence due to the compaction of the stratiform alluvion which constitutes the base on which the above sea-level hinterland and lagoon islands rest) and eustasy (fluctuations in the average sea level). The immediate form and the evolutionary trends of a lagoon depend on the relative importance of the above-mentioned phenomena and, in particular, on their temporal fluctuations on geological, climatic or short-term time-scales, induced by human or natural intervention. The evolutionary trends in relation to a particular hydrodynamic regime depend on transport mechanisms: the net transport of sediments to the lagoon channels; the net transport of sediments from the channels to the sea and possibly from the sea to the inside of the lagoon; the complex interaction between the transport mechanisms at the sea/lagoon interface; the introduction of sediments of fluvial origin; the artificial removal of sediments through excavating and dredging.

The technical problem of the dynamic equilibrium relationship between lagoon form and hydrodynamics is complex (O'Brien, 1969; Bruun, 1978; Jarret, 1976; Marchi, 1990; Di Silvio, 1992). In the first place, the notion of equilibrium has to be understood as the balance of sediment production and transport: in a confined environment such as the lagoon from which the rivers carrying sediments with their floods and silting have been excluded, an 'equilibrium' cannot exist unless it is one of zero transport. Or rather, either the lagoon is hydrodynamically dead since it cannot transport material even only suspended by internal wave action, or the erosive activity cannot be considered balanced since it lacks the entering loads.[52] Moreover, estimated soil production and accretion rates cannot balance seabound transport (Day *et al.*, 2000). Therefore the hydrodynamic regime which is intended to be given to the lagoon only decides the timing of the complete demolition of the lagunar forms.

The transport capacity of the lagoon channels depends (Rinaldo *et al.*, 1999), on the water volume typically found in conditions which would dictate the maintenance of an adequate liquid section. The commonly used empirical relationship the liquid section of a stable lagoon canal (not in rapid evolution) with the average tidal volume exchanged during the year, known as the Jarret relationship (Jarret, 1976). The volume of the tide replaces the instantaneous maximum volume which, in turn, determines the maximum transport capacity. Some caution should undoubtedly be adopted in the general use of such a model:

(i) the comparison between tidal inlets and internal lagoon channels is not an obvious one, since it must characterise the important diversity in the nature and the spatial distribution of the sediments on the bottom, and of the loads of sediments of coastal origin or coming from the areas adjacent to the channels;[53]
(ii) the morphological signs of climate change are generally present and visible and they are displayed in the expansion or contraction of the channel network (Rinaldo *et al.*, 1995) and in the concave areas (*i ciari*) from which the first incisions originate;
(iii) the transport mechanism which discriminates between the formation and the propagation of the incisions which form the network of channels in order to branch subsequently and the uniform flattening out of the shallow areas is not generally known.

Therefore, with the present state of knowledge, it cannot be said with certainty whether a modification of the hydrodynamic regime can create a network of lagoon canals or not. It is, however, reasonable to imagine that a minimum threshold for hydrodynamic forces at the bottom can be estimated under which sediments are not mobilised (Marchi, 1990), and which therefore stabilises the bottom in relation to certain tidal events. In the case of the Venetian lagoon morphology, it was agreed to assess the nature of the equilibrium in relation to the volumes exchanged (and therefore maximum capacity) for tidal events characterised by maximum gradients of 26 cm/hour with a frequency of 1% (Consorzio Venezia Nuova, 1989). The results of the comparison between sections in 1970 and sections sampled in 1990 show that the model simplified à la Jarret correctly highlights the evolutionary trends of the single sections, but is obviously unable to identify the overall trend of the system which would require more detailed morphological models (Consorzio Venezia Nuova, 1997). In any case, the simplified method's capacities for prediction, precisely because of its simplicity, are remarkable.

The twentieth century construction of the large navigable channels (concluded in 1968) considerably accelerated the demolition of the remaining lagoon environment, inducing, with the increased vivacity of the tidal currents and the consequent capacity for sediment transport, a significant deepening of the shallows along the *Canale dei Petroli* and the drastic flattening out of the lagoon beds, the disappearance of the *ghebbi* and the demolition of the large areas of salt marsh.[54] Moreover, a significant sedimentation was caused in the natural channels of the Malamocco basin (the Spignon, Fisolo, Melison and other minor channels) since the overall effect of increased dynamics is not necessarily a deeper incision of the lagoon canals, but also the flattening out and the deepening (other causes of which are subsidence and eustasy) of the shallows and the filling in of the bordering channels. The comparison between the bathymetric readings of 1970 and of 1992 is distressing since it confirms the demolition of the lagoon's morphological structure in an irreversible way (in the sense of a natural process). There are no doubts, neither empirical nor theoretical, on the nature and extent of the morphological modifications caused by the *Canale dei Petroli.* Indeed, the process of transforming the middle lagoon into an open reach has been accelerated enormously by the action of modern man.

Some circumstances must be understood for the times in which the venture was conceived and for its advocates of that time, either because of the non-existence of models for predicting the morphological responses of the lagoon, or because the navigable channels were envisaged as the instrument for the revival of the city's port activity and, with it, the restoration of Venice's role of a great emporium of traffic and commerce that had made it so great in previous centuries. Furthermore, the *Canale dei Petroli* guaranteed the exclusion of oil tanker transport from the St. Mark's basin. It was, therefore, only the last of a long series of interventions aimed at the greatness of the city.

The Venice municipality has asked, even recently, whether some antidote to the degradation of the lagoon morphology can be found,[55] imagining it possible to ensure otherwise the transit of ships with a draught of up to 33 feet (i.e., by eliminating the straight tract of the Malamocco-Marghera channel) by restoring the navigability of the Fisolo Canal. This is a recurring theme: the desire for a prosperous city, an active port and a flourishing economy, and also an uncontaminated and natural environment. It is unlikely that natural tidal channels, which are not dredged and maintained and have meandering courses, are compatible with modern navigation.[56] If the latter is to be maintained out of social and economic considerations, the suitable lagoon morphology must be precisely planned and maintained over the years with large economic contributions and with little regard for spontaneous evolutionary processes.[57]

Finally, it should be observed that it is commonly held that the extent of the lagoon surface area and its depths univocally and simply determine the volume of the tide exchanged between sea and lagoon.[58] In reality, the morphological and hydrodynamic question is more complex. The variations in the depth of the incisions and shallows modify the resistance to action considerably. The consequences of this on the propagation of the tides and, therefore, on the morphology of the tidal network are by no means negligible. Furthermore, the variations in the surface area of the lagoon basin determine the variations in the typical time of the propagation of the tides. When the propagation time becomes comparable to the duration of the incoming tide, the flows going out interact with those entering, thereby causing an overall reduction in incoming tide volume in the lagoon. It has been observed (Di Silvio, 1992) that an optimal lagoon breadth exists (measured from the inlet to the boundaries along the principal channels with fixed dimensions) from the point of view of volume exchanged, and that the Venice Lagoon approximates such values within this configuration. It would therefore appear that the dimensions determined for the lagoon by the lagoon boundaries of 1791 are close to those which guarantee the maximum section of the inlets. However, it would perhaps be excessive to attribute such a providential choice to the wisdom of the Serenissima Repubblica's governors of the water. Rather, it is likely that the boundaries simply fixed the relatively stable situation which the lagoon had already reached after the river diversions of the two previous centuries. In any case, it is evident that the morphological analysis of the lagoon must not be restricted to the study of an improbable condition of equilibrium which is to be reached in geological time, but must be conducted by means of evolutionary models which are characterised by long timescales and are able to distinguish transient landforms from the long-term trends.

7. Altimetrical recovery

In my opinion, the *Insulae* project (Consorzio Venezia Nuova, 1989), should be considered very carefully. In fact, to make provision for the local defence of the centre and the islands by 'raising' the level of the city in a way that is permanent as well as compatible with the altimetrical, architectural and environmental conditions of the city has both historical roots and practical sense. There are certainly some problems for carrying it out. These result both from the different altimetrical positions of the various low-lying areas to be defended and from the irregular spatial distribution of the environmental and architectonic goods to be conserved. Indeed, the lowest-lying *insula* (that of St. Mark's) is also the one which must be treated with the most care because of its incalculable historic and artistic value. The establishment of the level of the defence elevation (the level below which there is no flooding in the project area) depends on many local factors and enables the prudent and comprehensive restoration of the connective tissue of the city. However, such a project must be conducted, as in the past, to counter, with comparable speed, the reduction in elevation caused by the combined effect of subsidence and eustasy (Datei, 1997). However, it also takes on particular vitality and importance if combined with the possibility of control with the closure of the inlets. Indeed, it is pointless to imagine a limitless elevation of the city without protecting it from the events, which are increasingly frequent due to the above-mentioned reasons, over about 1.00 m a.s.l.(present), a reasonable estimation of the technical elevation for protection of the *insulae*.[59]

8. Lagoon and ecosystem survival (water quality and morphodynamics)

The key aspect that needs be emphasised is the total dependence of the residual lagoonal facies and landforms on man-made interventions throughout the centuries.

As recalled above, the concept of natural equilibrium does not have any practical or conceptual application in the evolutionary dynamics of lagoons: in particular, this obviously applies to the evolutionary history of the hundreds of natural lagoons once disseminated along the northern Adriatic seashore now long disappeared, naturally and irreversibly. The very process of lagoon creation hinges on the unstable balance between land-borne production of sediments and their sea-bound transport induced by: the interactions with sea and tidal currents; resuspensions; depositions; and physical and biological transformations. Without artificial man-made interventions this lagoon would have disapperared, filled by riverine sediments, about 500 years ago. It was the strategic interest of the city (as a port, as a lively gate to the Eastern markets, and as a well defensible dwelling) that sustained a long and strenuous opposition to natural evolution.

Thus lagoon and ecosystem survival need the man's design. By design I mean a process of creation of widespread social agreement among the parts (in modern democracies, luckily, a few illuminated minds cannot deliberate on crucial decisions). This in turn requires the preliminary definition of sustainable development, both economic, urban and environmental, in a context opposed to natural trends. It seems too much to ask Venetians of the day, none of whom seems to have the stature of their predecessors as far as courage and drive are concerned. The farsightedness of the old rulers is controversial, and we may discuss the issue at length. In fact, many technical errors bearing major impacts were made through the centuries, largely because of an inadequate technical and scientific equipment. Today the lagoon morphology and ecosystem tend steadily to become typical of a sea branch, as suggested for instance by space/time distributions of mean salinity or a comparison of the accurate bathymetries of 1970 and 1990, as a consequence of a specific design choice made more than 400 years ago.

A current opinion, largely reflecting the loud conservationists' position, proposes that the main priority of intervention, rather than facing the flooding problem, is to consolidate and preserve the lagoon that was functionally operating in the mid-1850s, before the major intervention aimed at reshaping the northern and central mouths that regulated the sea–lagoon exchange. Large-scale conservation would mean a fairly active role in supplying suitable sediment. In fact, marshlands and tidal flats have been lost to an impressive pace during this century, largely due, rather than to more aggressive tidal currents as a result of evolving, sea-bound lagoonal facies, to subsidence and eustatism. It is to be remarked that some residual lagoonal landforms are still characterized by a extraordinary beauty and scientific interest and constitute an ecosystem of the greatest value. Once the elevation of a marshland falls below the threshold of, say, 0.10 m a.m.s.l., the mean submergence exceeds the survival capabilities of typical interface vegetation (typically *Spartina Leyantha* or *Salicornia*), and the erodibility of the zone increases dramatically leading to demolition by wave and current actions. It is clear, much to our despair, that unless major artificial interventions occur, chiefly aimed at supplying sediments reclaiming elevations lost to the sea, all residual landforms are bound to disappear within a few hundred years at the most. It is also clear, even out of simple comparison of relevant time and space scales in the suitable flow equations, that no morphological intervention short of closure may significantly affect the propagation of exceptionally high tides within the lagoon. This holds regardless of the expandability of the tides on large areas devoted to fish-nursery and farming, which would have a significant reduction effect for the city only for spring tides whose tidal gradients are large, notwithstanding widespread prejudices.

From the general mass conservation viewpoint, no balance of production of transport of sediments may occur in a context, like the one we experience today, in which all major tributaries have been diverted to preserve the sea characters. Supply from the outer sea and soil production due to accretion processes are negligible, and wind-driven resuspensions are major landforming processes even in the presence of gentle tidal slushing. Incidentally, this is the basic reason why no depth reduction of either the inlets of the inner channels can prevent erosion or even reverse current evolutionary trends in the large-scale morphodynamics. So if the question posed in the title of this paper, will Venice survive?, is to be referred to the residual Venetian lagoon the answer is no, unless we decide so. Natural evolution would liken the lagoon to a shapeless bowl in much less than geological times, just wait and see.

A considerable effort and a large amount of money has already been devoted to improving lagoon water quality by reducing pollutant and nutrient loads. Nevertheless, water quality is still poor and there are severe problems associated with blooms of the macroalgae *Ulva*. It has been noted that, since 1985, biochemical oxygen demand (BOD) loads to the lagoon have been reduced by 70%, but the nitrogen load is still high. Several future strategies have been proposed as necessary to produce a significant nitrogen load reduction, these include facilities to reduce the nitrate loads from the drainage basin as well as sewage collection and treatment for the city of Venice, construction of translagoon pipes and ocean outfalls. With regard to macroalgae, eutrophication of the lagoon has been shown to produce a build-up of organic matter in the sediment and to induce anoxia and hydrogen sulfide production in wide areas of the lagoon – *Ulva* beds may cover a third of the lagoon – standing crops may reach 450,000 tons of wet biomass and a yearly production of 2,000,000 tons.

Temporary closures of the mouths (designed to prevent, say, tidal elevations higher than 1.20 m a.m.s.l.) would not have any immediate significant negative impact on the lagoon water quality. However, in the future as sea level continues to rise and the frequency and duration of gate closures increase, the level of impact of the gates on water quality will depend on the success of parallel measures for major nutrient load reductions.

Planning for future lagoon nutrient load reductions under realistic economic and technological constraints must continue in parallel with the necessary flood protection works. The necessary tools are monitoring and modeling of the water quality and eutrophication processes that are unique to the lagoon. Models have been developed that have helped to answer important questions about the health of the lagoon. However, because of the scarcity of monitoring data necessary for model calibration and verification, these models lack the ability to differentiate between controlling nitrate from drainage basin runoff and ammonia nitrogen from the untreated wastewater of Venice, an important issue because of the known preference of *Ulva* for ammonia. Further theoretical and experimental studies are nonetheless required. Fortunately there seems to be enough time and will to undertake an integrated monitoring and modeling program designed to provide the information necessary to assign priorities for future pollutant and nutrient control facilities in the lagoon. Thus a healthy ecosystem seems bound to survive.

9. Which destiny?

The social and economic fate of the city is inseparable from the strategic choices for its protection. Indeed, no protection plan may exist which would not take account (more or less explicitly) of the prefigured use of the city. Preservation cannot also avoid the issue of what use will be made of the city and its environment by future generations. For example, a city with an active port wants (would like: if this were not considered a crime)[60] the division of the lagoon into separate tidal subbasins, perhaps featuring some form of storm surge protections for Venice and Chioggia and the acceptance of the middle lagoon's status of a sea arm. Revamping of the once world-famous port activities would indeed require a rather different setting of the mid-lagoon, with continuous access and channel depths that are doubtfully suited to other uses of the lagoon and the sustainable development of the current ecosystems. Similarly, the fate of industrial zones in the Venetian mainland and surrounding areas hinges on environmental impacts that are unlikely to be accepted, either now or in the future. But then one wonders that the middle lagoon has already been devoured by the *Canale dei Petroli* and thus depths suitable to modern navigation needs would perhaps be no matter of major concern. Moreover, the basin could be effectively utilised for diverting the undesired freshwater runoff from its drainage basin.[61]

The exercise of unconstrained planning soon turns rather futile. Besides, the subdivision of the lagoon into separate basins is probably not the best solution to the city's and its lagoon's current of future problems. It is inadmissible, however, though quite common to resort to bogus arguments such as, for example, those for the unity or the physical equilibrium of the lagoon, in order to maintain the unfeasibility of lagoonal partitions (born, in fact, out of a dintinguished Dutch engineer's mind in the aftermath of the 1966 disaster).

It is also obvious that the oil-tanker port is perceived as the deadly enemy of the lagoon environment (Rosa Salva, 1974, 1983). Personally, I believe that if the price to pay for social consensus on the conservation of the city is the sacrifice of the industrial port, then so be it (regardless of how much past rulers of Venice, who always maintained and tampered with the lagoon in order to preserve the port vocation of the city, might turn in their graves). But nonetheless it does not seem necessary to warn of a radical incompatibility existing between the conservation of the historical city and lagoon and the continuation of its important use as a port and industrial settlement, which would amount to a radical and sectarian ultimatum.[62] Widespread action for stitching back together and carefully reclaiming the low-lying areas of the historical centre and of the islands, together with the control of the exceptional high tides by means of installations at the inlets, would permit the conservation of the city along with tolerable interference in its port and industrial functions.[63]

The effects of the excavations of the large channels and the nineteenth century intervention have certainly been devastating for the morphology of the middle lagoon, now reduced to a shapeless hollow. These works were, however, aimed at giving Venice back a strong port vocation and at turning it back into a centre of traffic and commerce. In this way, they favoured the demands of the city over those of its (rather artificial) environment, although in mitigating cultural circumstances as a result of the incapacity, typical of the times, to clearly predict the effects and the transformations which intervention would have caused. Furthermore, the fault of the dredgings and excavations in this century is to have accelerated, indeed not outrightly generated, a natural process which was already begun with the diversion of the rivers which had created the lagoon with their flood-borne mud and sediment loads. Today, the port's problems are of a political and economic nature and are not necessarily related to the protection of the city. In exactly the same way, the passage from heavy industry to thoughtful industry[64] in Porto Marghera (a charming slogan from the Venetian Green Party) is irrelevant, in my opinion, for the key issues relevant to the conservation of the city and its lagoon.

In conclusion, the infinite *querelle* over causes and effects of *acqua alta*, the flooding, in Venice seem to be quite a bogus problem. Indeed, it could be considered resolved, since the distance between conflicting results is very limited if read carefully. Rather, it seems necessary for a sort of international peer review process to be introduced for any subject which is deemed controversial, in order to involve disciplines and opinions which have not been worn by decades of polemics[65]. In short:

(i) Exceptional storm surges can only be contained by temporarily isolating the city from the sea. The main cause of the increased, observed frequency of high tides flooding the city is due to the combined effects of subsidence and eustatism, which jointly claimed approximately 23 cm in this century. In a critically low-lying area like the Venetian coastal, the most likely scenario of sea-level rise would lead to the loss of the city within a hundred years, a fate which seems unacceptable to us and, surprisingly, not perceived as the highest priority. A corollary of mean sea-level rise is also most worrying. It is the foreseeable increased frequency and intensity of storm surgesrelated to mean sea level rise (the so-called climate tropicalization due to increasing temperatures and the greenhouse effect). Hundred-year return period storm surges, whose destructive power evokes the dreaded flood of 1966, will be shortening their interarrival time. To protect from such events no measure but the temporary closure of the mouths can be effective;

(ii) The nineteenth century interventions and those of this century have caused modest, yet significant increases in the amplitude of the high tides and notable morphological alterations. Diffuse morphological interventions may re-establish the lagoonal facies of the last century, including a reduced flooding rate from tidal events, at the cost of heavy works and a continuous maintenance. In no way, however, they can reverse the erosive trend that leads to the evolutionary shaping into a sea branch. I can figure valid arguments for intervening in the lagoon by building and maintaining wetlands, but strongly object at their representation as a solution to the serious flooding risk of the region. Reducing the cross sections of inlets is a futile attempt to affect morphodynamic trends that current mathematical models view as strongly negatively impacting from the morphological viewpoint, not to mention the detrimental effect that would be borne on navigation and water quality. In no case, in fact, sediment can be prevented from being mobilized, the main resuspension pump being wind-driven rather than tidal current-driven. In essence, further deepening of shallow areas and the filling of the canals would be worsened rather being improved by diffuse interventions. Their environmental and economic impacts are thus significant, especially within a context of imperfect management of pollution abatement, and their potential for protection from flooding from major tides is irrelevant;

(iii) The work of widespread reconstruction of the height of the historical centre is feasible and opportune. However, the combined effects of eustasy (the sea-level rise) and subsidence predicted for the next century threaten the city's destiny like Damocle's sword. The capillary work of reclamation of lowlands within the city can and should be done, up to elevations of the order of 11 m above mean sea level, possibly slightly more, at increasing costs for the architectural texture of the city. This is nevertheless a measure bound to be marginally significant unless coupled with protection from major floodings. Without them, the race to higher elevations by reclamation seems rather futile;

(iv) No major hydrological problem of freshwater inflow prevents extended periods of interruption of sea-lagoon exchange. Even unrealistically high scenarios for intensity and duration of freshwater inflows support that the threat of flooding from freshwater is negligible. Any detailed hydrological study cannot but reduce the related risks.

All that now remains is to assign priorities to the crisis factors that need be addressed, thereby assuming full political and historical responsibility for the course of events awaiting us. In the meantime, a type of tattered Disneyland[66] is inevitably asserting itself as the model for the development of Venice, as a result of the lack of strong and far-sighted projects, widely agreed upon and backed by broad support,[67] on the city's vocation for the third millennium. With respect to this model, I fear its superficiality, its materialistic drive (Venice's true, giant business) and the merely skin-deep image of security and cleanliness it wants to sell us.

Will Venice survive? It is impossible that the intertwined ecosystem that includes the city and its environment can survive as it is now, largely because it is the artificial byproduct of coevolution of human interventions and natural tendencies adapting to social or economic needs now long gone. History teaches that throughout the centuries the needs of the city always prevailed, but this seems to have faded from common and political perception. It is the city's future that matters, in my opinion. Indeed, the lagoon takes on vitality and relevance only because of the presence of the city in it. We must remind the worshipers of the lagoon *com'era dov'era* (as it was and where it was) that it has been continuously reorganised throughout the centuries in order to adjust the requirements for Venice's life and prosperity.[68] It is therefore an obvious conclusion that Venice is disappearing rapidly, in the physical sense and socially, and that the present generation of Venetians (because our responsibility is the greatest) will not be able to create the conditions for handing the city down to our descendants. But should the fate of Venice, the treasure of arts and the gift to the humanity, belong only to this generation of Venetians? Or maybe to the whole world and the future generations?[69]

Notes

1. The general need for a technically irreproachable basis for taking strategic decisions on the great themes of environmental protection was recently highlighted in relation to the destiny of the oil platform Brent Spar and the opposing technical evaluations of Shell Co. and Greenpeace, e.g. Nature, 382, 103, (1996); 381, 358, (1996); 381, 99, (1996); 380, 13, (1996).
2. C. Datei (1997). It is no coincidence that the author is one of the scholars who have most contributed to understanding the complexity of the hydraulic problem of the propogation of tides, cfr. e.g. Datei (1972).
3. There is little value in appealing to the far-sightedness of the Serenissima Repubblica (the 'Most Serene' republic) over the centuries, which, especially in hydraulic matters, was often forced to resort to hasty modifications of long-planned projects and works.
4. On the contrary, it is perhaps a unique case, see Morandini (1960).
5. See P. Bak (1996). Bak, whose name is linked to the theory of self-organising criticality (the theory claiming to explain the dynamic origin of the invariable forms of scale in nature), has devoted significant attention to this theme.
6. The celebrated Darwinian definition taken up by Jacques Monod in defining chance and

necessity as the motors of biological evolution.

7. It is no coincidence that, from a morphological point of view, some statistical for the tidal networks of the Venetian Lagoon are indistinguishable from those observed in other lagoons, which differs in terms of tides, geology, latitude, climate, etc. This is still more evident and documented for natural fluvial networks, for which a significant body of theory and observation exists, (see Rodriguez-Iturbe and Rinaldo, 1997).
8. The solution of the city's problems, and the possible adoption of controversial resolutions, requires widespread understanding of the Venetian reality between government authorities, citizens, and Italian and international public opinion. In recent years, numerous initiatives have served to aid (but sometimes also to mislead) aquisition of knowledge of a complex system such as that of the Venetian lagoon. In my view, the most modern, for the tools employed and the far-sightedness of the underlying idea, is the availability on the Internet of the database of the *Sistema Lagunare Veneziano*. This is an initiative of the Istituto Veneto di Scienze, Lettere ed Arti in the framework of wide collaboration between the Ministry for Further Education and for Scientific and Technological Research, various universities and the National Research Council. This free internet service is offered to the citizen with the sole aim of informing and providing data. The database can be accessed at the site http://www.ivsla.unive.it
9. I use the term hydrogeological order, regime/equilibrium, or imbalance, with reluctance because, although it is meaningful and widely used, it is inaccurate. Hydrogeological refers to the movement of water in geological formations, and therefore refers to subterranean phenomena. It is difficult to convey to foreign scholars, whose languages are employed in a less off-hand manner, that this term is used to refer to the ensemble of hydraulic and morphological phenomena which regard the physical environment of Venice. And for the student of hydraulic phenomena, the implicit oxymoron in the title assigned to me sticks out like a sore thumb, because physical, or hydrogeological, equilibrium cannot exist for a lagoon.
10. I quote *verbatim* some considerations of Dorigo (1973), a master of lucidity and method in the Venetian question.
11. See e.g. Zille (1957); Ghetti (1990); Leonardi (1987); and especially the fundamental Dorigo (1983).
12. On the subject of historical mishaps, the rapid abandonment of its bed by the mouth of the Brenta in the South Lagoon is recorded in a drawing by Paleocapa.
13. The reading of the historic record and the memorable debate between Sabbadino and Cornaro (see Cessi, 1930) lends itself to various interpretations with regard to the analogy between the great historical decisions of the sixteenth century and those of the present. On this subject, Adami and Baschieri (1995) is of interest.
14. For an interesting comparison, see the contrasting theories of Pirazzoli (1987) and Di Silvio (1992).
15. A quantity rather different from the physical velocity of tidal currents. This is the velocity at which a state of motion is propagated, e.g. a determined level. On a calm water body of depth h, for example, a small (with respect to h) intumescence (of the sort produced by throwing a stone into the water) propagates in all directions with celerity $\sqrt{gh}$, where g is the acceleration of gravity. In a still body of water of depth 2 m, the intumescence travels at a rate equal to $\sqrt{2 \times 9.81} = 4.43$ m/s as long as the velocity of the fluid is zero. This simple example should illustrate that the propagation regards the state of the liquid body and not the transport of the fluid mass, which is still. Incidentally, the correct expression of the rate for non-small wave disturbances in the presence of resistance to motion does not significantly differ from the former relation. This important difference between the physical velocity of currents (which can be directly observed), and the celerity of propagation of a wave (which cannot be observed locally and which causes the delay of the tide), illustrates one of most common technical misunderstandings in the field of lagoon hydraulics.
16. This circumstance was verified experimentally on the construction of the great barrage dam across the Zuiderzee, in Holland (1933) which reduced the surface area of that lagoon by almost one-third (Zitman *et al.*, 1991; Terwindt and Battjes, 1991). The theoretical interpretation of this phenomenon, suggested by Lorentz, is discussed for example in Di

Silvio (1992) and is briefly discussed in what follows. This helps to show that the Venetian case and the validity of approximate empirical laws are by no means general.

17. Of which little is known in reality, see note 60. Furthermore, models of density currents, generated by the failure of fresh and salt waters to mix and studying the glow currents so induced, are irrelevant for the purpose of predicting tidal peaks. They become relevant in considering water quality, as briefly discussed elsewhere in these notes.
18. Today, the state of the art numerical models available are of the finite elements and finite differences types. This is terminology which refers to the way in which discretization occurs and, therefore, the way of resolving the complex mathematics which defines the problem.
19. It is worth recalling the development through time of a hierarchy of mathematical models of the hydrodynamic behaviour of the Venetian Lagoon, cited in detail by Ghetti (1990): (a) the first static model cited above (1970), which assumes the instantaneous parallelism of the water body throughout the whole lagoon; (b) the first propagatory models along the channel network (Di Silvio and D'Alpaos, 1972; Goldmann *et al.*, 1975); (c) the first quasi-bidimensional approach, that is, capable of more sophisticated treatment of the interaction between propagation in the deep tidal channels and adjacent shoals, salt marshes and shallows which store water (cf. the description in D'Alpaos, 1992); (d) the present-day completely bidimensional numerical models (or even tridimensional in some applications) of high geometrical precision, up to the current complete model made up of 12209 elements (see Consorzio Venezia Nuova, 1989). The limits and validity of mathematical predictions and the accuracy of the numerical procedures have been discussed in depth, with the conclusion that the predictions of the mathematical hydrodynamic models are absolutely reliable (cf. Datei, 1992).
20. The sources of the experimental data are the tide measurements of the historical *Ufficio Idrografico del Magistrato alle Acque di Venezia*, the efficient records of the *Ufficio Maree del Comune di Venezia* and the experimental work of the *Consorzio Venezia Nuova* (1989).
21. Compare the theories set out in the text of the preliminary project *Riequilibrio e Ambiente* (Consorzio Venezia Nuova, 1989) with, for example, Rosa Salva, (1974, 1983, 1997); Rosa Salva and Rabagliati, 1978.
22. In reality, as I discuss below, there are no differences in kind between interpretations of the diagnosis of morphological damage produced by the navigable channels. If anything, it is the prognosis which is different.
23. e.g. Pirazzoli (1987). The certainties pronounced on this point are surprising: from Indro Montanelli (I refer to the press campaign waged from the columns of the Corriere della Sera from 1968 to 1973, and especially to the famous testimony during the defamation trial brought against him by the Mayor of Venice (1972), as denounced from Dorigo, 1972), to Bettin (1993), the lack of critical advice and scientific support in verifying opposing hypotheses on such important themes is inexplicable. The only possible explanation, although it is not a flattering one, is the view that hydraulic science was subjected to the culture, and the ensuing disasters, of *what is done*. In some cases this is probably true, but that does not justify the inexcusible technical superficiality. Simple, immediate explanations which are profoundly erroneous, such as the image of the fist in the basin to explain the relationship between lagoon surface area and the amplitude of the propagation of the tides, have led to confusion. As mentioned above, the notion of the propagation of a long wave, such as that of a tide, cannot be immediately grasped. It is described with rhetorical figures ('... growing masses of water pushed more quickly...') but obeys the laws of physics, the balances of mass and forces, which cannot be evaded.
24. e.g. Di Silvio (1992).Note that the amplification of the amplitude of an oscillation does not imply an equal increase in the maximum level reached in the lagoon. In fact the trough (i.e. the minimum level during the oscillation) of the tide is also modified. According to a linear model of the phenomenon, the increase in the maximum level should be equal to half the amplification. In reality, experimental observations of the tides measured in 1930 by Magrini made simultaneously at Punta di Lido and at Punta della Salute, compared to analagous present-day experimental records, show that the greater steepness acquired by tide waves

entering today causes a decrease in the minimum point. The consequent reduction in the maximum value of the oscillation is equal to the amplification in the amplitude (Cecconi, 1997, personal communication).

25. On this point the learned debates of the modern era can also be noted, which are moreover instructive since they convey a certain echo of the scientific value of the work. Supino (1970a; 1970b; 1970c) maintains that a model lagoon not very dissimilar to the middle lagoon after the construction of the *Canale dei Petroli* led to significant amplification of the tide (an average increase in level equal to 8%). This study gave rise to a current of opinion, still alive today, which holds that the channel caused the increase in the frequency of high water events in the city. This opinion was held in particular by I. Montanelli (who defined Supino as, the only honest person in the technical debate over the destiny of the lagoon; on this subject see the trial deposition cited in Dorigo, 1972). The study contradicted other studies of similar model lagoons, and was therefore subjected to close examination (Datei, 1972; Datei and Dronkers, 1972): with the conclusion that the discrepancies were clearly attributable to the incorrect method employed by Supino in setting background conditions for the model of tidal propagation. The true amplification was considered as significantly lower for geometric relations (this is a rectilinear lagoon model, broken by a deep central channel flanked by shallow, symmetrical lateral expansions) and depths close to those really existing for the Venetian lagoon. Moreover, the geometrical schematisation studied by Supino, Datei and Dronkers, unavoidable at that time due to the need to work with analytical models for solving the equations of motion, did not permit the direct extension of the preceding results to the case of the Venetian lagoon. The necessary generalisation could only be obtained some years later with the next generation of numerical models which are now an indispensable engineering tool (Di Silvio and D'Alpaos, 1972; D'Alpaos and Degan, 1977; D'Alpaos, 1982; Umgiesser *et al.*, 1988; Danish Hydraulic Institute, 1989b). While this list is very incomplete, it shows the progress of technical research over the years, to be considered as in place for applications at the end of the 1980s. In addition, in the particular case of the Venetian Lagoon, the role of resistance to motion cannot be considered negligible and this creates theoretical problems due to the nonlinearity of processes of this kind. Also of interest is the consideration of Di Silvio on the Lagoon's inherent characteristics of being close to hydraulic resonance, which justify small amplifications in the tide waves.
26. This will be examined below.
27. One important characteristic of the tides, with reference to their propagatory aspects, is the average gradient, defined by the relationship between the total range of movement performed by the water level and the period of time over which this range takes place. This gradient, generally expressed in cm/h, summarises information both on the amplitude of the oscillation and on its period and can be statistically adjusted. A typical syzygial tide with a maximum sea level equal to 90 cm a.s.l. has an average gradient of 9 cm/h which takes place (on average) 19 times a year. The maximum level produced at Punta della Salute by all mathematical models is of 93.5 cm a.s.l. ± 1 cm depending on the various hypotheses of resistance to motion. A tide of average gradient equal to 13 cm/h, with a maximum sea level of 90 cm a.s.l., has an average frequency of 13 times a year and produces a maximum level at Punta della Salute of 93 cm a.s.l. The tide of 4 November 1966 (average gradient 8 cm/h) produces no amplification. In any case, the amplification is clearly dependent on the particular characteristics of each tidal oscillation. An average of tidal gradient weighted for frequency of occurence gives a total amplification effect of 2/3 cm more than in the conditions prior to the excavation of the great navigable channels (D'Alpaos, 1992; Consorzio Venezia Nuova, 1997).
28. It has already been noted that for every tidal oscillation in the Adriatic (characterised by an amplitude and a period, or combination of these) and for every internal point of observation there is a specific corresponding amplification/reduction of the propagation of the wave in the lagoon. The 'average' amplification therefore depends on the relative frequency of high tides generated in the sea. For the present configuration, compared with the configuration prior to the excavation of the navigable channels, an average amplification of 2–3 cm has been estimated (see Note 26). On the other hand, the condition of the nineteenth-century lagoon was very different. In fact the reduced dimension of the tidal openings dissipated *ab*

initio a great part of the wave's energy, while propagation within the lagoon, characterised by internal canal depths of the order of 3–4 m, was dominated by hydraulic resistance phenomena. The morphology of the nineteenth-century tidal networks was responsible for the substantial reduction of tides characterised by large average gradients, while at the same being unable to defend the city from exceptional events for which propagation characteristics become irrelevant due to the persistence of anomalous levels at sea.

29. See the following discussion.
30. In particular, if analysis with moving averages is extended not only to the maximum value (maximum annual level), but also to the average of the first 5 or 10 maximum values, the amplification is much smaller (equal to around 5 and 3 cm, respectively) (Scotti, 1997, personal communication. Also in Consorzio Venezia Nuova, 1997). This result indicates that the different amplification effect of different tides can also be objectively identified in the statistics. This result also tends to give greater validity to the estimates weighted for frequency of events obtained by mathematical models. Furthermore, in comparing maximum values recorded by the tidal gauges at the Lido and at Punta della Salute, it is not clear if the data refers rigorously to the same tides, a necessary condition for comparison of any significance. Finally it is interesting to observe that analysis with moving averages conducted on tidal minima show no increasing trend, and the analagous analysis carried out by Pirazzoli for the tidal gauge of Trieste also shows, curiously, an increase of around 10 cm in 100 years, which can certainly not be imputed to the excavation of navigable channels at a distance of hundreds of kilometers (Consorzio Venezia Nuova, 1997). All this gives good cause to doubt the validity of the statistical procedure.
31. Considerable literature exists on this specific problem. In particular a great contribution of ideas and culture focusing on the debate over the physical preservation of the city can be found in the proceedings of the Istituto Veneto di Scienze, Lettere ed Arti (1961–1996). Specific analyses of tidal propagation in conditions of various reclamation works and for different dimensions and geometry of incisions dictated by the tidal channel network are in the volume uniting important contributions on the significance of the lagunar boundary in the proceedings of the aforementioned conference organised by the Istituto Veneto di Scienze, Lettere ed Arti di Venezia, (see Adami, 1992; D'Alpaos, 1992; Datei, 1992; Di Silvio, 1992; Ghetti, 1990; Maccagni, 1992). Works of great technical interest include the studies of the Danish Hydraulic Institute (1989a) and the synthesis in Consorzio Venezia Nuova (1989). Finally the irreproachable technical discussion of Dorigo (1972, 1973) must be mentioned for its historical importance.
32. The issue of the probable frequency of exceptional events is complex and cannot be dealt with here. I shall state only that the common estimate which gives the event of 4.11.66 a return time of around 200 years is unconvincing (Ghetti, 1973; Pirazzoli, 1991; Consorzio Venezia Nuova, 1989). In fact the statistical adjustment of the Gumbel distribution to the sample appears inadequate in all the treatments which I have been able to consult, apart from being technically unsatisfactory in some ways. Self-organising phenomena tend to follow the Pareto or Lèvy distributions, which have algebraic tails which do not decay exponentially (cf. Bak, 1996). As a result, extreme events will be far more probable in reality than they appear to be from the distributions of the Gumbel type.
33. There is not much literature on this subject (cf. P. Rosa Salva, *op. cit.*). There is however a copious collection of public intervention and a specific request from Venice Town Council to the Consorzio Venezia Nuova (March 1995).
34. This last measure also figures, with explicit reference, in the laws for Venice (in both 171/1973 and 798/1984).
35. The so-called bent breakwaters (Calligaro, 1988), a technical and practical absurdity, cannot be considered as a real proposal.
36. On the other hand I am unable to evaluate the socio-economic and political implications of these proposals, in particular with regard to the economic survival of the port function of the city with reduced depths. The central question seems to me to be related to the destiny of oil tanker traffic and the petrochemical terminal at Port Marghera, which, however much it may

inspire dislike, provides work for 5000 people. The maritime traffic statistics which I have superficially consulted appear to show that the great bulk of commercial traffic could be sustained even with depths of 9–10 m, but in my opinion the conditions which might permit the development of a modern port function with the survival of the lagunar ecosystem have yet to be clearly identified. On this subject the reader is directed to other sources (Consorzio Venezia Nuova, 1989; Rosa Salva, 1983, 1997; Musu, 1998).

37. The lack of an adequate cultural forum for the political decisions regarding protection has distant origins, and is principally due to the aforementioned, presumed subservience of science to the culture of 'what is done' and the instrumental use of technical arguments. Transparency of communication is not aided by the fact that, during debates, the attention of technical opinion usually appears to be circumscribed, laying itself open to charges of narrow-mindedness. However, in view of the complexity of the problems to be addressed, this is perhaps the only way to avoid indulging in facile superficiality.

38. One estimate (from studies of the concessionaire Consorzio Venezia Nuova, 1997) of the volumes of bed material needed to reduce depths and to eliminate the *Canale dei Petroli* is of 7.2 million cubic metres (to which around 5.6 million cubic metres must be added for the coastal channel), a significant quantity (and cost). Even leaving aside the question of the stability of the bed once put in place, considerable expenditure can be envisaged (comparable with the cost of work at the inlets) even if demolition of the *casse di colmata* (reclaimed areas) is employed to procure the material. I remind that the *casse di colmata* were in fact reclaimed with the material from the excavation of the Malamocco-Marghera channel (1968) (Rusconi, 1987). It must also be noted that the recalibration of the Spignon and Fisolo, the corollary of this intervention, would require the excavation of around 6 million cubic metres of sediment and continuous maintenance afterwards. On the other hand, what cannot be envisaged is restoring the 23 cm lost by the lagunar bathymetry this century as a result of subsidence, which would require around 40 million cubic metres of bed material.

39. On this subject see the aforementioned contributions to the volume Conterminazione Lagunare: storia, ingegneria, politica e diritto nella Laguna di Venezia (Istituto Veneto di Scienze, Lettere ed Arti, 1992).

40. On this subject see the study with various mathematical models commissioned by the Consorzio Venezia Nuova (1989, 1997). The estimate of the reduction in average-high tidal peaks which can be obtained by using *all* the proposed wide area measures together is little more than 5 cm. Specifically, the effect of all the wide area measures on the most common syzygial tides (90 cm a.s.l. as the maximum level in the sea, average gradient equal to 9 cm/h, probable frequency 19 times/year) is of –5.1 cm; for more steep tides of maximum level 90 cm (13 cm/h) the reduction effect is of –8.2 cm. For less steep tides the reduction effect rapidly tends to zero. For tides characteristed by a maximum level at sea equal to 140 cm a.s.l. (for which the mechanisms of resistance to the motion of propagation are modified in ways which are neither trivial nor negligible) and a gradient of 10 cm/h, the total reduction is of around 6 cm. Significantly, the reduction of the peak is less than 2 cm for the tide of 4.11.66 which, as is known, peaked at 1.94 m over the zero at Punta della Salute (around +1.71 over the current average sea-level), with an average gradient equal to 8 cm/h and the abnormal duration of around 20 h. But even these negligible figures are misleading, because for around 16 h the level in the sea and in the lagoon would coincide in the region of 1.50–1.70 m a.s.l. The reduction of 2 cm would be obtained because the peak which reached 1.94 in 1966, starting from a level around +1.70, was relatively fast. These evaluations can certainly be considered precise and are clearly indicative, in my opinion, of the scant hydraulic relevance of wide-area measures, even if taken to the point of restoring the lagunar morphology of the early nineteenth century: in a historically questionable process and with significant worsening of the technical, economic and environmental problems described here.

41. Cf. note 52. It should be noted that tidal volumes expand more quickly in a lagunar basin along the deepest incisions, and propagate themselves in the adjacent shallows more slowly. As growing volumes penetrate the lagoon, the increasingly more internal channels gradually swell with sediments mobilised from the shallows (which are therefore deepened) which are

generally deposited in the principal channels. The principal tendency of every new incision is to be filled as a result of this mechanism. The restoration of the lagunar forms proposed by wide area measures can not eliminate the erosion induced in the shallows by the wind-driven re-suspension even for the smallest tidal currents. This suggests the re-adaptation of lagunar morphology to changing input volumes (a process which in 30 years has now passed its acute phase) which has as its corollary a further deepening of the shallows and the need for patient maintenance and excavation work on the new Spignon and Fisolo channels. In fact dynamic equilibrium of production and transport of sediment can not be obtained – it is impossible to fix it to the bed – without the production of sediment. The destiny of the lagunar morphology needed to be taken into consideration earlier (I mean around 400 years ago). But even if we could decide to redirect the rivers into the lagoon we could not be certain of the survival of the lagunar geomorphology. In fact, the great filter of fluvial sediments constituted by the dams of Cadore and Trentino would have to be dismantled, in this way eliminating a great deal of hydroelectric generation and the significant contribution to the hydraulic security of the North East territories made by the flood-control reservoirs. In addition, a large part of the mountain territories would have to be abandoned to permit the reforming of the large alluvial cones which constituted the reserve for transport by the historical floods. Whether it was right or wrong, the decision of the Serenissima Repubblica cannot be altered today. The destiny of the lagoon, or its degradation in the absence of any intervention, remains to be established. Human intervention can only slow down or accelerate the work of demolishing the residual lagunar forms. The demolition of the shallows caused by the combined action of erosion and subsidence, and greatly accelerated by the excavation of the lagunar channels, cannot reasonably be compensated for by inflows from outside and cannot be stopped (or even reversed) with wide-area measures. The acute phase of demolition in response to the hydrodynamic modifications carried out at the end of the 1960s has now been finished. Such a situation is known in geomorphology as the tendency of the topography of a mature profile towards a peneplain in areas free from tectonic activity: since the transport of sediment (not compensated for by a rise in ground level) is generated by topographic gradients, this will cease asymptotically only when all gradients are eliminated. Mature fluvial landscapes, geologically old, are gently undulating and not very 'deep'. Thus, as a result of the processes underway, the lagunar form, to which the present environment is irreversibly tending, is a reach without internal incisions and of little depth: our only influence over this is in determining whether this occurs over a geological time-span (which is not easy to predict) or over the span of a few generations.

42. The theme of the quality of lagunar waters, today and in the future, is only touched on in amarginal fashion in the report, which focuses on a different kind of equilibrium. The state of knowledge of the lagunar ecosystem and the reliability of mid- and long-term predictions must be considered inadequate on the whole. The hostility of environmentalists to the so-called dilution of polluting loads is certainly understandable, a non-solution of environmental problems. However, the total clean-up of the Venetian lagoon is a utopian ideal, in my opinion, for technical reasons connected in particular to the role and nature of pollutants from diffuse sources (transported, that is, by waste water of atmospheric origin) for which traditional purification processes are not practicable. My pessimism is compounded by the accretion of competence over environmental matters (state, regional, provincial and/or municipal), the huge underlying business of anti-pollution measures and the continuing absence of any real authority in the condition to take decisions on a technical basis and in a transparent fashion. It therefore does not seem very likely that there can be any certainty in this matter after decades of failed general regulatory plans for the clean-up of the lagoon: the only certainty is that if the lagoon were not properly cleaned up and depths were reduced, it would rapidly become a sewer. It is difficult to imagine a decision-making process which found this situation acceptable: a balance sheet with sure losses and very uncertain profits.
43. Aggravated by the suspicion that political hostility is the product of easy demagoguery for their ownership by propertied interests and their usage. Certainly the opening of the valleys would imply compromising their present production activity at no hydraulic advantage.

44. '... An enormous mass of water was therefore repulsed from the valleys towards the city, and also from the filled and industrial areas. The pressure on Venice therefore increased immeasurably...', (Bettin, 1993).
45. For a collection of varied reference material see Consorzio Venezia Nuova (1989).
46. The untenable assumption that there are structural measures, such as the widespread reduction of depths, which are capable of protecting against exceptional high water events, has already been noted.
47. The relationship between subsidence and extraction from the subsoil has only recently been established theoretically. In the case of Venice, the Venetian CNR Institute has played a key role in the study of subsidence since it was founded, (Tomasin, 1992; Gambolati and Gatto, 1975; Carbognin *et al.*, 1981; Carbognin and Taroni, 1996b). Important theoretical results are also to be found in Gambolati *et al.*, (1974a, b).
48. There is also a vast literature on this specific theme. For interesting recent results see Carbognin and Taroni (1996) on tide measurements over the last century at Punta della Salute (unstable for vertical translations) and Trieste (stable) whose contrasts reveal the effect of subsidence.
49. An important recent comparison of the various hypotheses can be found in Schneider (1997). There are various possible scenarios: (i) the disastrous melting of the polar ice cap (West Antarctica ice sheet), which would increase global sea-level by several metres; (ii) the continuation of the general eustatic trend of the last 20 years (after many technical and theoretical difficulties, determined to be +2 mm/year, mainly on the basis of measurements made by the TOPEX/Poseidon satellite); (iii) the possibility of irregular increases, suggested by experimental observation (from deep ice core samples or from deep samples of oceanic sediments from which surrogates of average annual temperature can be measured) according to which over geological eras eustasy has never manifested itself as a regular trend, but rather as a highly irregular fluctuation, with enormous changes occurring rapidly after long or very long periods of stasis. The comparison is worrying for the probable extent of the phenomenon, estimated to be from +20 to +50 cm over 100 years, and for the capacity of local governments to take advance action against the consequences in coastal areas ('... Responsible international authorities must take the full range of possibilities into account in planning for the future...').
50. A recent reference text is International Panel on Climate Change (1995). The balance of the available evidence, while allowing for local variations, suggests that human activity has a tangible influence on the climate, predicting an increase in the average global temperature of between 1.5° to 4°C by the year 2100. Despite the significant uncertainties involved in making forecasts, this should give cause for alarm, especially because such precautionary estimates indicate large increases even without invoking disastrous phenomena such as the melting of the Antarctic ice caps. The city would therefore disappear in the course of a century.
51. Pirazzoli (1991). It should be further noted (Schneider, 1997) that another concern related to the increase in average temperature is the progressive tropicalisation of the climate, which implies an increasing frequency of high-intensity atmospheric events. These, caused by higher thermic gradients, generate strong local anomalies, violent disturbances and exacerbated meteorological tides (storm surges). For example, it is precisely the increase in storm tides and maximum, rather than average, levels which most worry Dutch engineers.
52. With the possible exception of the contribution of sediments brought to the lagoon via the inlets as a result of interaction with littoral transport processes. This may be of some local significance in the vicinity of the inlets, but in general I hold it to be irrelevant for the morphology of the lagoon as a whole.
53. For which significant spatial heterogeneity is to be expected even only as a result of different conditions of vegetation with depth (in shoals and salt marshes; or in relation to the relative prevalence of fresh or salt water).
54. See for example the Table comparing bathymetries from 1970 and 1993 in Adami and Baschieri (1995).

55. Comune di Venezia, Note presented to the Inter-ministerial committee on 15 March 1995.
56. See the considerations stated in response to the queries of the Municipal Administration in Consorzio Venezia Nuova (1997).
57. The considerable quantity of material needed to fill in the *Canale dei Petroli*, and the problems likely to be encountered as a result of this, have already been remarked on.
58. I have discussed the problematic and complex relations between maximum tide levels and lagoon surface area elsewhere in these notes. An interesting systematic study (Consorzio Venezia Nuova, 1997) on the propagation of tides of amplitude equal to 1.40 m and variable period clearly demonstrates that the change in conditions between 1930 and the present led to only a marginal increase in the amplification of lagoon levels (on average 5–6 cm out of 140), but a significant increase (up to 63% for tides of periods equal to 6 h) in the instantaneous maximum flows carried by the inlets. It must be repeated that it is not licit to correlate maximum flows with maximum levels in the lagoon. The greatest instantaneous flows have rapidly caused significant morphological changes without, however, dramatically altering the situation with regard to maximum levels, which have only increased marginally.
59. At up to the level of 1.00 m a.s.l. (in 1976), around 4% of the surface area of the historical centre was affected. Up to 1.20 m, 35%, and over 1.40 practically the whole city was flooded (Frassetto, 1976). The present situation is only marginally different (Consorzio Venezia Nuova, 1989). It is clear, however, that a limitless altimetrical recovery through raised grounds is not conceivable for it entails destruction, rather treasured conservation, of the city's priceless architecture.
60. The physical unity and indivisibility of the lagoon are explicitly referred to in the aforementioned Laws for Venice, 171/1973 and 798/1984. I am not at all convinced that cast-in-stone principles of unity and indivisibility exist for a lagoon which, until the end of the last century, for example, possessed five inlets.
61. The current estimate of the contribution at full flow of the drainage basin of 1000 m^3/s (Cavazzoni, 1973) is improbable from the hydrological point of view for the method and hypotheses with which it is calculated. But as often happens, a figure which is little more than an exercise becomes the supporting evidence for discussion of, for example, the highly improbable flooding of the closed lagoon due to freshwater discharged from its drainage basin: the contribution at full flow is then extended in invalid fashion to the whole duration of a possible operation of closure of the inlets. In any case, the need for an in-depth hydrological study of the contribution of fresh water on a solid and scientific, rather than schematic, basis is absolutely clear. This is also the case in other environmental issues not considered in this note.
62. I quote verbatim Dorigo (1973), whose lucidity and rigour has already been noted.
63. At least for the next, critical, 100 years. One hundred years is in fact the projected life-span for mobile (rather than fixed) installations at the inlets; this also represents the time lapse within which, with all probability, the combined effects of eustasy and subsidence (and of the increase in frequency of extreme tidal events) would make any port activity prohibitive due to the continual opening and closing. At that point, the necessary and painful choice between the city and lagoon would have to be faced.
64. A well known, and rather hard to translate from italian, slogan: '*Dall'industria pesante all'industria pensante*', which rhymes nicely.
65. But peers it must be, in the sense of the cultural tradition of the enlightment, so far removed from the prevailing paradigm that treats scientific knowledges as all equal, and, of course, equally irrelevant.
66. 'The emphasis shifts towards superficiality, tending to a Disneyland-like artifact…' (Munk and Munk, 1972).
67. I am deeply convinced by the validity of Bettin's analysis on this theme.
68. I quote verbatim a phrase of Ghetti (1984, 1985) which has always struck me.
69. I owe several acknowledgements, *in primis* to Claudio Datei, my guide and reference on Venetian matters. Wladimiro Dorigo has always been a source of inspiration for personal and cultural reasons. The work of, and the interactions with, Paolo Costa and Ignazio Musu have

been very influential. Everybody lectures on hydrodynamics in Venice. It is the climate, I suppose, yet some Venetians taught me well, the late Augusto Ghetti, Aldo Rinaldo, Alessandro Franchini and Paolo Tito. I borrowed the subtitle of this Chapter from Donald R.F. Harleman who coined it for the proposal of a conference. I also benefited from my interactions with colleagues at MIT: Rafael L. Bras, Paola Malanotte Rizzoli, Chiang C. Mei, Donald R.F. Harleman.

Bibliography

Adami, A. (1992) Alcuni aspetti idrodinamici legati all'apertura delle valli da pesca rilevati su modello fisico della laguna. In: *Conterminazione Lagunare. Storia, ingegneria, politica e diritto nella Laguna di Venezia*, Venice, pp. 293–306.

Adami, A. and Baschieri P. (1995) Gran porto fa gran laguna. In: *Atti e Memorie*. Venice, IVSLA, Vol. 153, pp. 48–56.

Ammerman, A.J. and McClennen, C.E. (2000) Saving Venice, *Science*, 289, 1301–1302.

Ammerman, A.J. *et al.* (1999) Sea-level change and the archaeology of early Venice. *Antiquity*, 73, 303–312.

Bak, P. (1996) *How Nature Works. The Science of Self-Organized Criticality*. New York: Copernicus-Springer.

Bettin, G. (1993) *Dove volano i leoni*. Milan: Garzanti.

Bruun, P. (1978) *Stability of Tidal Inlets*. New York: Elsevier.

Calligaro, M. (1988) Un piano per l'Area Lagunare Veneziana. In: *L'altra Venezia a Convegno*. Venice, pp. 14–20.

Carbognin, L., Gatto, P. and Mozzi, G. (1981) La riduzione altimetrica del territorio veneziano e le sue cause. In: *Rapporti e Studi*, Venice: IVSLA, Vol. VIII, pp. 55–83.

Carbognin, L. and Taroni, G. (1996a) Eustatismo a Venezia e Trieste nell'ultimo secolo. *IVSLA Proceedings*, CLIV, 281–298.

Carbognin, L. and Taroni, G. (1996b) Linearità fra due variabili: piezometria e subsidenza nell'area veneziana. *IVSLA Proceedings*, CLIV, 33–52.

Cavazzoni, S. (1973) Acque dolci nella laguna di Venezia. In: *Laboratorio per lo studio della Dinamica delle grandi masse by CNR, Rapporto Tecnico n. 64*.

Cessi, R. (1930) *Antichi scrittori di Idraulica Veneta* reprinted by the Ufficio Idrografico del Magistrato alle Acque of Venice, 1987.

Chastel, A. (1990) *L'arcipelago di San Marco*. Venice.

Consorzio Venezia Nuova (1989) *REA: Progetto preliminare di massima delle opere alle bocche*, Vol.1–4.

Consorzio Venezia Nuova (1997) *Studio di Impatto Ambientale (SIA) del progetto di massima della chiusura mobile delle bocche di porto*, Preprint Technical Report.

Co.Ri.La (Consorzio per la Gestione del Coordinamento Attività di Ricerca Inerenti de Sistema Lagunare Veneziano, Venice) (1999) Scenari di rescita del livello del mare per la laguna di Venezia, pp. 1–40.

D'Alpaos, L. (1982) Correnti residue fisiche e numeriche nei modelli di propagazione unidimensionali. In: Proceedings of *Processi di mescolamento in ambienti lagunari o estuarini e loro modellazione*, Rome, pp. 79–93.

D'Alpaos, L. (1992) Evoluzione morfologica della laguna di Venezia dal tempo del Denaix ad oggi e sue conseguenze sul regime idrodinamico. In: *Conterminazione lagunare. Storia, ingegneria, politica e diritto nella Laguna di Venezia*. Venice, IVSLA, pp. 327–358.

D'Alpaos, L. and Degan, F. (1977) Un modello matematico Lagrangiano per lo studio del movimento d'acqua nella Laguna di Venezia I:, *Rapporti e Studi*, Venice: IVSLA, Vol. VIII, pp. 148–160.

D'Alpaos, L. and Di Silvio, G. (1975) *Studio su modello matematico degli effetti idrodinamici che conseguirebbero all'apertura delle valli da pesca nella Laguna di Venezia*. Padova: Istituto di Idraulica, University of Padova, Studi e Ricerche, n. 293.

Danish Hydraulic Institute (1989a) Inlet stability and impact of new works. In: *Rapporto Tecnico B.7.3*, Venice: Consorzio Venezia Nuova.
Danish Hydraulic Institute (1989b) Modello idrodinamico bidimensionale di dettaglio della bocca di Malamocco. In: *Convention n.6479, Technical Report,* Venice: Consorzio Venezia Nuova.
Datei, C. (1972) Sulla propagazione della marea in una laguna schematica secondo l'impostazione bidimensionale. *Istituto Veneto di Scienze, Lettere ed Arti, Rapporti e Studi*, V, 43–80.
Datei, C. (1992) Esame comparato delle caratteristiche dei modelli matematici della Laguna di Venezia. *Conterminazione lagunare. Storia, ingegneria, politica e diritto nella Laguna di Venezia.* Venice: IVSLA, pp. 271–291.
Datei, C. (1997) *The problems of Venice: an introduction*, ICOLD XXXII General Congress Proceedings, Florence, May, preprint.
Datei, C. and Dronkers, J.J. (1972) Considerazioni sulla nota del Prof. G. Supino La marea nella Laguna di Venezia. *IVSLA, Rapporti e Studi,* **VI**, 186–203.
Day, J.W. *et al.* (1999) Site accretionary dynamics, sea-level rise and the survival of wetlands in Venice lagoon: a field and modelling approach, *Estuarine, Coastal and Shelf Science*, 49.
Denaix, A. (1811) *Carta topografica idrografica militare della laguna di Venezia….*, Venice.
Di Silvio, G. (1992) Delimitazione topografica di una laguna a marea: implicazioni idrodinamiche e morfologiche. In: Conterminazione lagunare. Storia, ingegneria, politica e diritto nella Laguna di Venezia. Conference Proceedings. Venice: IVSLA, pp. 307–325.
Di Silvio, G. and D'Alpaos, L. (1972) Validità e limiti di un modello matematico unidimensionale nella propagazione di onde lunghe in campi di moto con caratteri parzialmente bidimensionali. In: *XII Conference Idraulica e Costruzioni Idrauliche proceedings*. Milan.
Dorigo, W. (1972) *Una laguna di chiacchiere. Note a margine a Tutto Montanelli su Venezia*, pro manuscript. Venice: Tipo-litografia Emiliana, pp. 120.
Dorigo, W. (1973) *Una legge contro Venezia. Natura, storia, interessi nella questione della città e della laguna*. Venice: Officina Edizioni.
Dorigo, W. (1983) *Venezia Origini*. Venice: Electa.
Dronkers, J.J. (1964) *Tidal Computations in River and Coastal Waters*. Amsterdam: North Holland.
Frassetto, R. (1976) Altimetria del Centro Storico di Venezia. *Boll. Geof. Teor. Appl.,* **18**, 3–16.
Gambolati, G. and Gatto, P. (1975) Simulazione della subsidenza di Venezia. In: *Venezia ed i problemi dell'ambiente*. Bologna: Il Mulino, 299–360.
Gambolati, G., Gatto, P. and Freeze, R.A. (1974a) Predictive simulation of the subsidence of Venice. *Science*, **183**, 1023–1024.
Gambolati, G., Gatto, P. and Freeze, R.A. (1974b) Mathematical simulation of the subsidence of Venice. *Water Resources Research*, **10**, 563–577.
Gatto, P. and Carbognin, L. (1981) The Lagoon of Venice: natural environmental trend and man-induced modification. *Hydrol. Sci. Bull.*, **26**, 379–391.
Ghetti, A. (1973) I problemi idraulici della Laguna di Venezia. *CNR Quaderni della Ricerca Scientifica*, **94**, Rome.
Ghetti, A. (1984a) La salvaguardia di Venezia. In: *Una cultura per Venezia*, Conference proceedings. Venice: IVSLA, pp. 9–23.
Ghetti, A. (1984b) *Riflessioni su Venezia espresse agli amici*, pro manuscript, Treviso: Grafiche Zoppelli, pp. 23.
Ghetti, A. (1990) Italian contributions to hydraulics of lagoons. *Excerpta,* **5**, 7–55.
Ghetti, A., D'Alpaos, L. and Dazzi, R. (1970) Il Comportamento della Laguna di Venezia indagato con il metodo statico. University of Padua Istituto di Idraulica, *Studi e Ricerche*, n.9.
Goldmann, A., Rabagliati, R. and Sguazzero, P. (1975) *Characteristics of the tidal wave in the Lagoon of Venice*. IBM Tech. Rep. CRV 009/513–3539.
International Panel on Climate Change (1995) *Second Assessment Report* (J. Houghton ed.), Cambridge: Cambridge University Press.
Ippen, A.T. (1996) *Estuary and Coastal Hydrodynamics.* New York: McGraw Hill.
Istituto di Idraulica G. Poleni, University of Padua (1970) *Effetti dell'apertura delle valli da pesca alla libera espansione della marea*, Technical Report.

Istituto di Idraulica G. Poleni, University of Padua (1979) *Le correnti di marea nella Laguna di Venezia*. Padua: Ministero dei Lavori Pubblici, pro manuscript, pp. 236.

Istituto Veneto di Scienze, Lettere ed Arti (1961–1996) *Rapporti e Studi, Commissione di Studio dei Provvedimenti per la Conservazione e Difesa della Laguna e la Città di Venezia*, Venice, Vol. I–XII.

Istituto Veneto di Scienze, Lettere ed Arti (1992) Conference proceedings: *Conterminazione Lagunare: storia, ingegneria, politica e diritto nella Laguna di Venezia*, Venice.

Jarret, J.T. (1976) Tidal prism-inlet area relationships. In: *G.I.T.I. Report no. 3*, Fort Belvoir, VA: US Army Corps of Engineers, Coastal Engineering Research Center, pp. 136.

Leonardi, P. (1987) Dati e problemi geoidrologici, meteorologici e paletnologici veneziani. In: *Proceedings of the Research Workshop A Vent'Anni dall'Evento di Marea del Novembre 1966*, Venice: IVSLA, pp. 7–54.

Maccagni, C. (1992) Motivazioni e criteri tecnici della conterminazione. In: *Conterminazione lagunare. Storia, ingegneria, politica e diritto nella Laguna di Venezia*. Conference proceedings. Venice: IVSLA, pp. 147–164.

Marchi, E. (1990) Sulla stabilità delle bocche lagunari a marea. *Rend. Fis. Acc. Lincei*, **9**, 137–150.

Morandini, G. (1960) Elementi geografici ed aspetti morfologici della laguna e della città di Venezia. In: *Conservazione e Difesa della Laguna e della Città di Venezia*, Conference proceedings. Venice: IVSLA, pp. 65–82.

Munk, J. and Munk, W. (1972) Venice Hologram. *Proceedings of the American Philosophical Society*, **116**, 415–442.

Musu, I. (1998) Venice and its Lagoon: a problem of local sustainable development, in this volume.

O'Brien, M.P. (1969) Equilibrium flow areas of inlets of sandy coasts. *ASCE J. Waterways Harb. Div.*, **95**, 1306–1315.

Pirazzoli, P.A. (1982) Maree estreme a Venezia (periodo 1872–1981). *Acqua-Aria*, **10**, 1023–1039.

Pirazzoli, P.A. (1987) Recent sea-level changes and related engineering problems in the lagoon of Venice. *Prog. Oceanogr.*, **18**, 323–346.

Pirazzoli, P.A. (1991) Possible defenses against a sea-level rise in the Venice area, Italy. *J. Coastal Res.*, **7**, 231–248.

Rinaldo, A. *et al.* (1995) Geomorphic signatures of varying climate. *Nature*, **374**, 632–635.

Rinaldo, A., *et al.* (1999) Tidal networks, 2 and 3. *Water Resources Research*, 35, 3905–3929.

Rodriguez-Iturbe, I. and Rinaldo, A. (1997) *Fractal River Basins: Chance and Self-organization*. Cambridge: Cambridge University Press.

Rosa Salva, P. (1974) Trasformazioni ambientali ed alterazioni nella laguna veneta. *Urbanistica*, **62**, 5–44.

Rosa Salva, P. (1983) Environmental management of the Venice Lagoon. In: *Proceedings IGU, Comm. Coastal Environ.*, Bologna, pp. 33–38.

Rosa Salva, P. and Rabagliati, R. (1978) Indagine di fattibilità di un modello per la pianificazione del territorio lagunare. *Acqua & Aria*, **9**, 697–709.

Rusconi, A. (1987) *Variazione delle Superfici Componenti il Bacino Lagunare*. Ministero dei Lavori Pubblici, Ufficio Idrografico del Magistrato alle Acque di Venezia, *Pubbl. n. 160*, Venice.

Sabbadino, C. (1930) *Scritture sopra la Laguna*. Reprinted in: R. Cessi, *Antichi scrittori di Idraulica veneta*, Part I. Venice.

Schneider, D. (1997) The rising seas. *Scientific American*, **3**, 112–118.

Supino, G. (1970a) *L'influenza di variazioni recenti di superficie e profondità sulla marea lagunare di Venezia*. Internal Report, Istituto di Idraulica, University of Bologna.

Supino, G. (1970b) *Sopra la marea nella laguna di Venezia – Complementi sperimentali*. Internal Report, Istituto di Idraulica, University of Bologna.

Supino, G. (1970c) The propagation of the tide inside a lagoon. *Meccanica*, **1**, 42–53.

Terwindt, J.H.J. and Battjes, J.A. (1991) Research on large scale coastal behaviour. In: *Proc. XXII Conf. on Coastal Engineering*, Delft, pp. 89–99.

Tomasin, A. (1974) Recent changes in the tidal regime in Venice. *Rivista Italiana di Geofisica*, **3**, 3–6.

Tomasin, A. (1992) L'attività scientifica sui problemi lagunari condotta dal Consiglio Nazionale delle Ricerche. In: *Conterminazione lagunare. Storia, ingegneria, politica e diritto nella Laguna di Venezia*. Conference proceedings, Venice: IVSLA, pp. 259–270.

Umgiesser, G., Sudermann, J. and Runca, E. (1988) *A semi-implicit finite element model for the lagoon of Venice*, Technical Report.
Zendrini, B. (1723) *Idrografia de' Porti e Fuose di S. Nicolò, di S. Erasmo, dei Tre Porti*, Venice.
Zille, G.G. (1957) L'ambiente naturale. *Storia di Venezia*, Vol. I, Venice.
Zitman, T.J., Stive, M.J.F. and Wierswa, J. (1991) Geological and historical development of the Dutch coast. In: *Proc. XXII Conf. on Coastal Engineering*, Delft, pp. 89–99.

CHAPTER 4

Urban Sustainability and Territorial Structure

Carlo Magnani and Traudy Pelzel

Part One
Territorial Structure: Specificity and Analytical Problems

Premise

The vast number of sustainability indicators which has emerged from recent experience does not take the form of a systematic framework which is methodologically coherent and universally shared in all its aspects. Nevertheless, the aims which are periodically expressed in the cases examined show significant convergence towards environmental issues and, furthermore, they attempt to set up a description of the parameters for a more general idea of quality of life as an objective to pursue in a medium- to long-term perspective.

The boundaries of the field concerning territorial structure, which define the scope of this work, delineate the field of thought within which to formulate the hypotheses for the construction of objectives which are qualitatively significant from the physical-morphological perspective. This produces questions, though mostly without answers, about a consolidated heritage of numerical indicators which is the product of decades of analytical and planning improvement in the land planning disciplines. Indeed, the physical structure of our cities, at least in their recent areas, is the product of numerical indicators which have not always proved capable of guaranteeing qualitatively significant results.

In taking on the overall picture of environmental issues, it must be noted that, whereas for some phenomena the objective of fixing a set of accepted numerical indicators constitutes the initial approach for a survey which is systematic and comparable to the phenomenon itself, for others, it is a matter of examining in greater depth a debate which might, perhaps, go beyond this methodological structure.

These difficulties are adopted as the centre of this research with the aim of constructing a process within which the consideration of indicators which are, so to speak 'discursive', graphic and numerical, advances at the same pace.

In particular, various keys to the processes of transformation of territorial phenomena are proposed as a premise to the formulation of lists of indicators.

I. Musu (ed.), Sustainable Venice, 95–120.

Thus, the aim is to create a sort of 'convention on meaning and terminology' for the specific indicators proposed for territorial structures which are listed together with other contributions in the general chapter on indicators, accepting a certain epistemological unreliability for the territorial planning disciplines as a background.

The concluding paragraph regards the formulation of various evolutionary scenarios for the Venetian settlement system. One of these is represented by the preliminary *Document alla Variante Generale del PRG* (preliminary document to the General Modification of the Town Planning Scheme) drawn up by the Municipality in conjunction with the development of this work and to which we have often referred throughout this research .

The practice of prefiguring scenarios is an attempt to describe potential situations by starting with the information collected and increasing some variables periodically. This also permits both taking on the difficulty of forecasting as one of the elements which characterises the current debate on the city as well as the examination of a particular settlement structure with the aim of producing potential capacities clearly, whether they be positive or negative, which this settlement structure can display or conceal.

1. Sustainability and physical-morphological aspects of an urban system: elements for a technically pertinent analysis

Generally, the study of an urban structure in terms of sustainable development aims to resolve an often conflictual relationship between development and environment. Consequently, the evaluation of a hypothesis of sustainability for such a structure involves tackling all those processes which relate to environmental protection in terms of the appropriate management of resources (water, air, land, green areas), a policy for the collection and disposal of rubbish, optimisation of the transport network, with the improvement of public services etc. Whilst accepting the general framework, the aim here is to contribute to the debate on the issue of introducing a qualitative objective for settlement forms, beginning with the consideration that the meaning of inconvenience which is increasingly often perceived in the contemporary urban space is also connected to the physical nature of places and to the dimensions of spaces. As a result, the design quality 'can help in raising the level of the work's environmental compatibility, thus contributing to make the pressure exerted by the work on the environment more sustainable as well as, therefore, by the city as a whole on it' (Karrer, 1993).

For this reason, the recurrent theme in the development of this work becomes that of the search for a pertinent expression of all of the aspects concerning the urban form; aspects which all too often, in previous experience, were described exclusively with the use of numerical expressions. This is done by evaluating the opportunity for putting forward a discursive model which, instead, could also take morphological, connecting and locating factors into consideration. The need to do so results from the fact that the use of a numerical indicator as an interpretative key for urban structures can display strong similarities with the town-planning practices inherited from the rationalist culture which confronted the issue of inequalities in supply (Secchi, 1994), approaching it in mainly quantitative terms. Following the crisis of such a culture of measure and of quantities placed throughout the territory with only distributive criteria and, as a result, of the failure in the idea of equilibrium as an equitable territorial distribution, the numerical indicator has shown characteristics of over summarising and of inefficiency in terms of inspection and control of urban quality in its entirety.

As mentioned above, several previous local Agenda 21 experiences contain very few examples of an evaluation of the anthropised environment which has been conducted by analysing the physical-morphological aspects of the territorial structures. Indeed, the lists of indicators which relate to the urban area take the functional aspects of the system into consideration almost exclusively rather than the physical ones.

1.1. Urban quality and sustainability

In recent years an increasing interest in that which exists has been witnessed and, consequently, a progressive shift in urban planning from projects for developing and expanding urban areas to projects for transforming the already urbanised territory has occurred. This shift in perspective entails a complete rethinking of the city and has the objective of increasing urban quality, whether by means of renewed attention to the physical dimension of the anthropised space rather than the purely quantitative dimension, or by going beyond the concept of areas which are homogeneous in terms of their function and, as a result, driving the different areas towards greater integration.

Although some authors believe the 'simultaneous maximisation of multiple objectives' (Munda, 1995) to be impossible, both a general idea of balanced interrelations between areas and the importance of maintaining a certain level of functional diversity as necessary conditions for a system's long-term existence constitute one of the possible interpretations of sustainability for settlement forms.

In terms of urban strategies this hints at the possibility of guaranteeing the system's capacity for self-regulation through time, ensuring a certain multi-functionality and re-balancing of parts of the territory by means of a precise system of relationships. In this way, over-simplifications of the urban structure which are the result of theories on specialisation and on functional separation are avoided. To evaluate the sustainability of an urban system can mean, therefore, considering 'a general problem of proportions in the intensity of use of the different parts of the territory and of the limits to urban capacity which these parts are designated' (Secchi, 1996a).

Having assumed the concept of sustainability in terms of balanced interrelations, of limits to development, of proportion and greater connectivity between the areas, a double series of indicators which would measure both the functionality of the urban system as well as its degree of connectivity and the dimensional relationship of the urban space can be foreseen as an objective of the research. Taken together, all these indicators contribute to the functionality of the system itself. As a result, such indicators will act as instruments for measuring the trends and progress towards an objective of sustainability undertaken as a politically significant aim rather than as a measure of particular threshold values which are difficult to quantify.

In this way, to reconsider the practices of urban design in terms of sustainability also means attempting to order the widespread settlement forms which lack inter-relations and synergy, and which, with constant land consumption, progressive impermeabilisation of soil and a continual extension of the infrastructure network, often constitute one of the main conflicts of interest between collective space and private space. Precisely for this reason, the affirmation of a new shared vision becomes increasingly urgent and necessary; an affirmation which, however, still comes up against the cultural problem of a poor education in the planning of a comprehensive transformation of the territory.

From this perspective, the three paragraphs on firstly, the idea of the city and its forms of representation, secondly, some cases of urban design as a form of public investment and, thirdly, the outlining of various phenomena of progressive inflexibility of parts of the urban area, are an initial attempt at reaching agreement on a common basis for dialogue. As a result, they must be understood as propaedeutic to the problem of the capacity to transmit the message. Indeed, the communication aspect emerges as being a central one to the entire Agenda 21 process. This issue is already of great importance in formulating the indicators themselves, as the example of Seattle clearly demonstrates (Lawrence, 1996). Here, a study was made which aimed at formulating indicators which could increase understanding of the problems. The reaction which this provoked amongst the population was so emotional that it caused a shift in policies and behaviour towards the change.

As a whole, the approach is that of examining scenarios which would permit the change in the individual actors' concepts of relative utility, thus prefiguring the shift of perspective from short-term to long-term. This appears more problematic in developed countries because of the need to combine the comparability of the results with the specificity of situations in relation to highly complex systems.

1.2. Sustainability in terms of performance evaluation

An evaluation in terms of sustainability can be envisaged as one in which the establishment of the reciprocal relationship between physical project and strategic project is crucial. In this perspective, such a correspondence can also be established in terms of performance by opportunely widening and examining such a concept in depth in order to be able to investigate the characteristics of the project, thereby verifying its adequacy and, consequently, its sustainability in relation to the pre-established objectives (Linch, 1996).

What is actually sought from this is to discover the reasons for the present impoverished spatial configuration which, although it maintains a high degree of coherence with contemporary demands and behaviour, ought to be reconsidered. This should not necessarily be done by beginning with a radical modification in such behaviour so much as with the promotion of a renewal which begins with the ways of settling and of relating, shifting attention onto the possible related transformations in space/landscapes; these last two are often treated in an exclusively technical way.

2. Boundaries of the field

It can be stated that it initially appears that to analyse territorial aspects related to changes in a settlement structure means focusing on the visibly, perceptively 'evident transformation', by considering the physical aspect of the transformation as the specific field of enquiry. This is considered in terms of its projection, such as the design techniques with their related forms of presentation and, therefore, that which can be communicated of the transformation's prospects before beginning work, as well as in terms of the physical outcome of the transformation, that is, the realisation of the project into something concrete.

A further specification of the problem is represented by the idea that the principles of the transformation which can be taken into consideration must be related to the transformation of that which exists in the sense that what already exists is, at least in dimensional terms, the broadly predominant part of what can be considered. As a result any model re-foundation in another place is, by definition, excluded.

This aspect, taken together with the halt in demographic pressure, at least in European countries, means that phenomena which have been witnessed in previous periods of expansion, that is, in the periods of urban agglomerations' spiralling growth through the large consumption of non-urbanised territory, must be considered to have finished.

From this perspective, non-urbanised space loses the characteristic of being land awaiting urbanisation, in accordance with a typical paradigm of the Italian Town Planning Scheme (PRG) and becomes, instead, a resource in itself in the proper sense within a broader vision of the anthropisation process.

Furthermore, a phase of city growth within the basic perspective of modernisation can also be considered almost over. This is a phase commonly bound to the execution of public works aimed at the satisfaction of primary needs (schools, roads, hospitals, etc.), which can be easily expressed according to quantitative parameters.

The consideration of urban quality depends on the comparison with all of the previous statements. Urban quality has taken on characteristics of a common ground, that is, of an inexorable objective of local government propositions and, at the same time, of a confused vagueness since a broadly shared cognitive heritage which has codified indicators of quality for a territorial structure in physical terms does not exist.

Periodically, reference can be made to the provision of equipment, the presence of services, and the functioning in terms of the relationship between space and time. In other words, reference is made to indicators which often tell of something other than the physical nature of the phenomena.

This difficulty is further increased both by the failure of the descriptive capacity of classic notions of opposition – city–countryside, centre–periphery – which, for a long time, have constituted the founding paradigms for the attribution of specific characteristics to certain areas of the territorial structure. In addition, the difficulty is increased by the renewed use of descriptions expressed in the form of opposing pairs based on the opposition of old-modern and constraint-project (transformation-conservation), which nourish forms of culture and opinion which are both nostalgic with respect to the unattainable quality of the old city and resigned towards the bad quality of the modern city.

The evident transformation of the contemporary city can only take place in discrete and limited parts which hint at the global without, however, ever being able to express it in a complete and synthetic form. Moreover, the Italian urban system and, in particular, the north-eastern one, is so dense and interrelated that it sometimes makes it difficult to understand the limits of one settlement structure in relation to another.

This suggests the formulation of the hypothesis that a distinction might be made between strategic intervention on the one hand, that is, that which hints at the global (whether in terms of opening up opportunities or of inhibiting them), with, obviously, the notion of global being subject to the relativity of the reference scale which is periodically employed, and, on the other, intervention for maintenance and widespread improvement, evidently trying to avoid thus acceding to, a new opposing pair based on the mutual exclusion of meanings.

Furthermore, a specific feature of territorial structures is their permanence in the long-term and the high level of inertia with which they counter the transformation processes. This means that the idea of the transformation must be considered in the medium to long term and the examination of the reactivity itself must contend with the difficulty of working with samples or with simulations.

The same decision-making mechanisms which govern a territorial structure's transformation involve the activation of procedures which are so demanding in relation to the time available that they sometimes place in doubt the cause–effect relationship between bringing the phenomenon to light, formulating the proposal for its solution and carrying out the work. Indeed, whilst a fiscal decree, for example, can be put into effect immediately, the same is not true for a decision-making process which aims at the transformation of a territorial structure. Moreover, within such a process, the notion of the point of no return does not exist; that is the notion of a time threshold beyond which the intervention programme becomes certain and can no longer be halted.

2.1. Value judgements: contradictions

All of the territorial planning instruments in Italy, from regional ones to provincial ones right down to those which fall within municipal competence, cover the whole territory in accordance with a conceptual procedure which goes from the general to the particular, without having, however, a required time connection.

Without entering into the town-planning debate or into the debate on the degree of coherence within the various levels of territorial planning, we can say it is not possible, or it is possible but only with a certain degree of caution, to talk of a settlement structure's spontaneous development.

This significant public investment constituted by town-planning, is often spent in stop-gap forms, that is, in the acquisition of short-term consensus (for example, for the duration of an administrative session), which is created around the most basic demands expressed by social groups and their simple summation, thus according to a logic of redistribution, rather than in the definition of a project's idea as a clearly oriented choice. Thus, a contradiction appears in the use of long-term materials (all that which concerns the hypotheses for the transformation of a territorial structure), which are used up, instead, in the short-term, with no checking for coherence. Doubts about the supposed spontaneity of certain forms of settlement structures arise from these contradictions. These points of view sometimes defend the chaotic complexity as being a literary metaphor for the incessant transformation. Such a vision would consequently entail giving up any form of presentation of a settlement's future through the town-plan.

However, it must not be forgotten that the product of these contradictions are settlement forms which put considerations on sustainability in a critical position in the sense that they grow with a constant consumption of the non-urbanised territory without establishing interrelations or synergy as well as with a progressive impermeability of soil, and with a continual need to increase their levels of infrastructure in order to satisfy primary needs, rather than to set development prospects in motion.

A further contradiction can be seen in the passage from 'what' to 'where and how'. For example, some territorial intervention sectors appear saturated as the biased opposition to the laying of new roads sometimes demonstrates, even when the functional need for them is generally accepted; or else in other cases, a confrontation occurs between local and global which is often able to halt the decision-making processes, as, for example, in the construction of water purification plants or waste disposal systems.

Paradoxically, the acquisition of some more advanced cultural levels of perception of territorial problems, such as of a more careful consideration of environmental problems, nourishes a description of the territory and of its characteristics which are considered as primary resources and not available for further evident transformations.

The transformation's uncertain prospects produce a user who is concerned only with their status as consumer of the final service, and who is, therefore, not willing, to take on complete responsibility for the long-term production cycle.

The ways of presenting the projection aspects of territorial phenomena (reports, plans and projects) often take on the form of promise whose fantastic content is expressed by the constant objective of 'qualitative improvement'. The inaccuracy of this notion, related to the difficulties of the Town Planning Schemes 'technical form' of expression (for example, the time limit set by a single threshold with respect to which it is rather difficult to establish priorities for execution) has made a series of criticisms possible both about the presentation's technical material and about the projection's ideological material. In short, the description of a future state can do nothing but become a story as soon as it is forced to abandon the paradigms of the scientifically exact forecast or the paradigms of a mechanical cause and effect relationship.

Despite the conviction to also work on ideological materials, nothing detracts from the importance of examining in-depth and with more precision and pertinence the form of promise, with the aim of promoting the creation of an active cultural climate which is able to transmit the meaning of the envisaged transformations' orientation and the meaning of the proposed objectives. Taken together, the promise and the rules for carrying it out appear as the basic conditions for being able to reactivate a collectively shared desire and to envisage a shift in perceptive capacities from short to long term, in order to be able to recreate hypotheses of re-appropriation for the future which are invested with planning capacity.

In this way, even the idea of preserving Venice is by no means an univocally fixed objective. The measures which aim at the survival of the city, such as, for example, the maintenance of the canals and the canal banks, merely establish a threshold of minimal rationality, the necessary condition for being able to put forward any other argument.

However, from the point of view of functional specialisation, the considerations which can be developed on Venice depend on the territorial reference scope.

On a large scale (even only that of north-eastern Italy), the functional specialisation of one part of the territory (Venice) does not figure as an element of crisis in the territorial system.

If, on the other hand, a more localised angle of vision is adopted, the over-specialisation in terms of tourism of such a considerable part of the territory takes on the form of an excessive inflexibility of the whole functional structure, thereby exposing the system to a clear fragility bound to its mono-function.

Without wishing to enter into ideas linked to the notion of equilibrium or re-equilibrium, in the sense of the fair and equivalent redistribution of functions or resources, it is possible to dwell upon the attempt to investigate those measures which allow the division of specialisation phenomena: on the one hand (the tourist phenomena) by redistributing them over a larger area with the aim of reducing existing levels of congestion, and on the other (the demographic and social phenomena), by investigating measures which might simply suggest the renewal of the demographic structure as well as the maintenance of a structure of consumed goods (in terms of typology and cost) and of ways of life which are as normal as possible.

A further contradiction can be seen in the low cultural profile which characterises some forms of participation in the decision-making processes with respect to the complexity of the problems. This directly corresponds to the often excessive complexity of the procedures and to the problem of relevant information.

In this sense, a monitoring programme should also be set up, for the territory's evident transformation, which would be able to reconstruct the entire framework of coherence between the different intervention scales, as well as the hierarchies and priorities.

In addition to being a programme of constant activity which promotes the co-ordination and efficacy of the measures, this ought to find its own place in which to take on concrete form through a permanent and constantly updateable exhibition of the planning documents and of the resulting projects as the main instrument for communicating the transformation processes. Above all, however, it would be necessary to set up participation programmes in the form of 'civic education' related to the problems concerning the transformation of the territorial structures with the purpose of facilitating an appraisal of the proposed technical solutions' relevance, as well as of the coherence of the investment programmes and of the physical features of the transformation hypotheses.

3. Interpretative hypotheses towards indicators

On the one hand, the numerical indicator has the property of representing a codification which utilises universally recognised terminology and, on the other, of often leading to mechanical transpositions between quantitative and qualitative aspects. In particular, the results of much of both the theoretical and practical tradition of town-planning, whether in its analytical or in its projection aspects, have found their parameters of reference in the machine-like ideology and one of their specific forms of expression in the numerical aspect. It can be sustained that the form of visible phenomena is often determined by numerical parameters produced by homologous indicators, which cover the territory entirely and which, though very morphogenetic, from an epistemological perspective have lost their meaningful link with the original problems.

Therefore, it is not so much a matter of taking a critical stance towards the numerical parameter since it probably remains a necessary form of expression, but, rather, one of understanding its aspects of inefficiency and partiality on the one hand and, on the other, of setting up an operation of reactivating its meaning.

In this way, it is possible to propose the attempt to establish by convention some issue-indices which could bring about the interaction of the description of the phenomenon with some broader levels of consideration so as to create some keys, a sort of discursive indicator, which might lead to the consideration of numerical indicators in a focused way.

The idea of the city and the form of its representation, morphological and functional inflexibility as result of a fragmented way of thinking and urban design as a planning technique capable of constituting a minimum threshold for environmental investment, become three starting points which make the attempt to go beyond the numerical indicator real.

3.1. The idea of the city and the form of its representation

The prefiguration of hypotheses for a territorial structure's planned transformation evokes, at least as a background, an idea of the city. Within this prefiguration, general theories intertwine which express the degree of total cultural progress in considerations on anthropisation processes as well as intervention techniques concerning the level of development of the disciplines pertaining to territorial issues.

The ways of applying the techniques, which constitute the main instrument for representing the idea of the city, identify several principles suitable for this specific case and thus establish a field of coherence according to which it is possible to speak of the pertinence of the planning techniques in relation to the proposed objectives. In this case, a consideration of planning techniques necessarily involves both an in-depth examination of the procedural aspects which can give concrete form to the hypotheses formulated, as well as a study of the existing resources. Here, resources must not be understood in an exclusively economic–financial way: the geographic and environmental structure of a settlement and the very stratification of the forms of anthropisation are in themselves resources, as long as investigation and interpretation can reveal propensities and capacities for transformation.

In conclusion, it can be maintained that just this moment of difficulty in formulating forecasts highlights to a greater extent how the physical-morphological structure of a settlement can take on the character of partial autonomy, thereby assuming the form of a supply in relation to which investments and transformation strategies should be directed.

For example, some episodes in the recent urban history of Venice can be read according to this key: from the G.A. Selva Town-Planning Scheme in the first years of the nineteenth century to the realisation of the Marghera working-class district (Fig. 1) which is a simplified differentiation of the idea of the garden city but whose basic design has acquired a permanent nature, being, still today, the effective example of functioning and of identity.

Or rather, more strictly concerning the urban part of the mainland, it is possible to refer to the first drafting of the Town-planning Scheme of 1937 and 1942 (Fig. 2), as far as to the realisation of the district of St. Giuliano, better known as Villaggio San Marco (Fig. 3), which includes plans for a considerable part of the territory between the Salso Canal and the river Osellino, thereby establishing a strong relationship between the city and the lagoon and, furthermore, to the competition for the St. Giuliano saltmarsh area (Fig. 4), where the winning project gave thorough form to the idea that the mainland urban centre could face towards the lagoon.

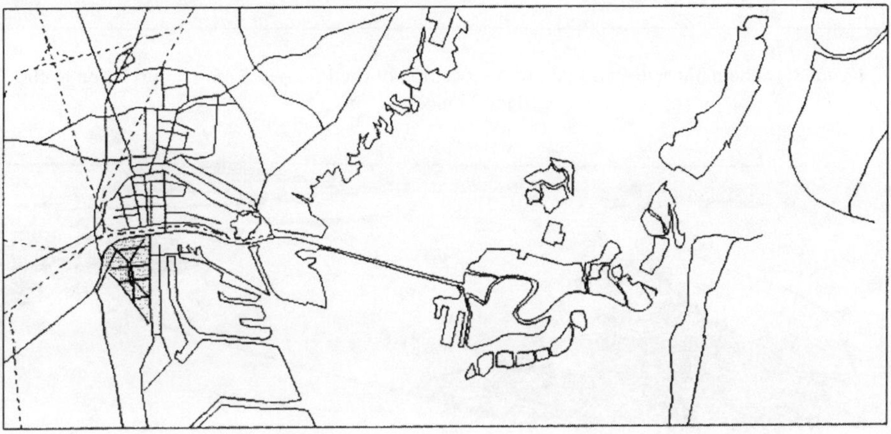

Figure 1 Schematic representation of the position of the Marghera working-class district in relation to the Island-Venice system.

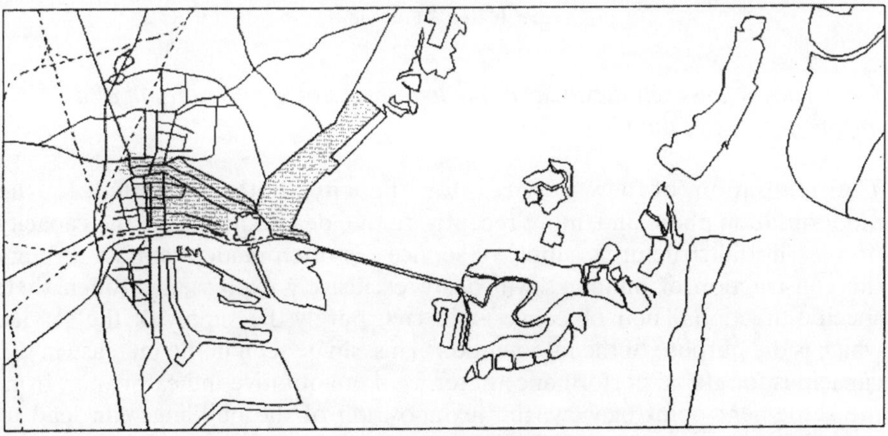

Figure 2 Schematic representation of the extent of the territory to which the 1942 Town Planning Scheme applied.

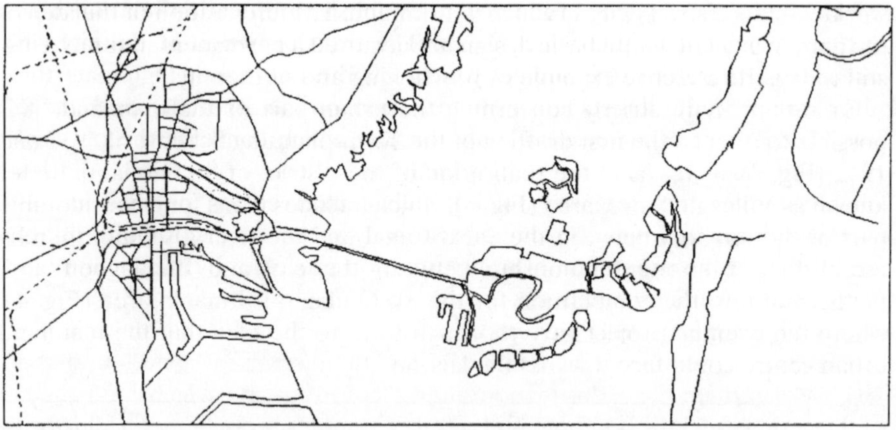

Figure 3 Schematic representation of the position of the *Villaggio S.Marco* in relation to the island-Venice system.

Figure 4 L. Quaroni and associates. Competition for the new district in the St. Giuliano salt marsh area. Aerial view.

3.2. *The progressive inflexibility of the territorial structure as a result of a fragmented way of thinking*

The realisation of new services for the city, both connected to the modernisation phase and, more recently, to the idea of increasing the capacity of a territorial structure's supply of services, is often made manifest through the construction of buildings which are exclusively expressed through their specific function. Their objective is, in fact, purely the supply of the service which is the purpose of the intervention. This can be seen in the increase in the capacities for global performance in terms of quantitative indicators but, from the settlement point of view, the juxtaposition of the buildings can lead to processes of progressive inflexibility in territorial structure. Indeed, the principle of juxtaposition can hardly be seen as the principle of stratification, the latter being the capacity of the new to establish modification relationships which are pertinent to that which exists. The transformation of the city (in existence) is, therefore, the subject which, on the one hand, highlights the limits of the procedural mechanisms in force and, on the other, demands more refined judgement capacities, not only for that which concerns the physical aspects, but also for the re-establishment of collective interests. All this probably implies that the very role of public intervention policies has to change: work can no longer be aimed exclusively at the satisfaction of primary needs, rather at the promotion of transformation processes by means of more complex and flexible procedures.

From this perspective, the physical character of the transformation itself, the relationships of measurement and scale, the dimensions of spaces and the relations of sequence, of disjunction and of co-ordination which bind them together constitute the fundamental issue. Examples from Venetian history can easily be found: from the construction of the railway bridge in the middle of the last century which cuts the lagoon and prevents the possibility to circumnavigate Venice (in the 1970s the widening of the railway bridge maintains the previous height above sea level) to the creation of the Stazione Marittima at the end of the century. In fact, the large reclamation works led to the absorption of the small island of St. Chiara which presided over the Grand Canal's mouth towards the land, thus powerfully intervening in the morphological redefinition of the whole urban area. In this way, the land-facing front takes on a form which is indifferent to the character of the city. The placement of the gasometer, of the water system and of a series of minor construction works have taken place with the same theory, embodying the idea of the modernisation of the city in a trivial and reductive way.

However, using this key, much more recent cases can also be analysed such as that, for example, of the urbanisation along the Romea Road behind Marghera (Fig. 5). Here, along the principal routes which lead into the urbanised countryside, heterogeneous buildings are often randomly clustered within their related undifferentiated open spaces, thereby forming placeless strips.

3.3. Urban design as a form of public investment

Urban design is a planning technique which permits the simulation of the physical outcome of programmes for the use and for the functioning of urban zones, by prefiguring the type of relations which can be established between the different elements (and, from here, go back to the specific operators) which contribute to the definition of a particular physical structure.

Whilst it is not univocally definable, the application's dimension is an important factor because it requires forms of expression which must consider functional synergy between the zones and what is built in them and, therefore from this point of view, which are forced to express cultural forms of a particular era inscribed in certain criteria of rationality. Given the durable

Figure 5 Marghera, commercial concentration along the Romea Road.

nature of settlement structures, this planning technique takes on the form of a kind of long-term investment. Indeed, it is believed that there is a value of operational and economic efficacy in what the urban design generates which, even if it is temporary value, must not be forgotten and, hence, a good urban design must also be able to withstand social marginalisation, to regulate the discrepancies between development and the existing, between tradition and renewal (Gregotti, 1986–1993).

What nourishes, perhaps, a certain nostalgia for the old city is the perception, beyond the visible form, of the cultural unit which permitted the consolidation of morphological structures which were so significant and, at the same time, so adaptable to changes in conditions of use. The continuity of an idea which is expressed in physically significant and perceptible terms is an expression of cultural continuity, even more so than one of political and social stability.

Bearing in mind the duration and the inert nature of the territorial structures, it is possible to arrive at the formulation of the paradoxical statement that an error pursued right to the end is better than continual changes of mind. In fact, the pursued error makes it possible not only to identify the phenomenon, but also to form the judgement as well as to introduce measures for its correction, as the case of Bijlmermeer, one of the districts of South Amsterdam, demonstrates. This case deals with a plan for restructuring a residential district with an excess of undifferentiated open spaces as its main problem; or in the cases of Orly and Venissieux which are proof of the French restructuring programmes for several large-scale residential intervention projects in which both town-planning and sociological issues are combined. Conversely, rhapsodic and fragmented projects create, instead, structures with are much more inert in nature and which are impervious to change.

What is more difficult in the matter of urban design is to limit the examples to the Venetian sphere, except by making reference to the urban district of Marghera and to Villaggio St. Marco again. These have been able to bring investments in building which have been made over the decades into line with general objectives for constructing a particular urban space.

European examples, on the other hand, are numerous and differentiated: from J. Nash's projects in London (1811–30) where the laying of a road and its shape form the link between St. James Park and Regents Park, to H.P. Berlage's proposals for South Amsterdam (1904–15) which produced one of the 'manifestos' for the continuity between town-planning and architecture, up to the reconstruction of Le Havre after the Second World War, where A. Perret posed the problem of identifying new rules, balanced between respect for the settlement tradition on the one hand, and for the city's renewal on the other.

If, however, the field were expanded to the relationship between built-up areas and green areas, it is possible to cite the case of Amsterdam, where the creation of the Amsterdam Wood was an objective pursued from the 1930s until the 1960s, or that of Copenhagen whose design is conceived as a continuous system of green spaces which connects the central parks of the city with the surrounding countryside, or the case of Stockholm, where the continuity of green within the urban scene is a structural element in the construction of the city.

Part Two
The Venetian Case: Evolutionary Scenarios

4. Scenarios

To propose a consideration by prefiguring scenarios means adopting a working method which is at the limits of ambiguity and based on exaggerating some of the basic facts. It also means, however, accepting confrontation with some of the difficulties outlined above.

In the first place, there is the difficulty of making forecasts simply on the basis of an analysis of the potential trend of numerical factors. Although there is no strong pressure for the growth of cities nor towards their generalised expansion, the procedural tools available still have the tendency to move in this direction. The system of the different operators' expectations, which essentially constitutes the general cultural climate of a city, is often orientated in the same direction, thus remaining anchored in the ideology of redistribution which has formed the essence of territorial policy for decades. The urban transformation programmes which are bound to specific situations and to the historical reality of individual cities and, which, out of necessity, have to concentrate investment-transformation opportunities in particular places are unlikely to suit simplistically redistributive reasoning. The urbanisation of agricultural land still appears to be a simpler and more obvious way of operating: indeed, it certainly proves to be more advantageous if one remains tied to a cost-benefit analysis which is limited to the individual project without a comprehensive analysis of the whole settlement and of its users.

On the other hand, the arguments in defence of environmental values, which are based on a purely conservative vision of the value of the existing, can do nothing but dwell on the objective of damage limitation, rather than on the construction of driving planning hypotheses. The redefinition of the collective interest, its division into partial and general objectives, as well as, consequently, the use of public policies for territorial management which can deal with these problems, constitute one of the limits which define the field and which require a radical change of direction in the idea of public investment. Faced with the difficulty of forecasting, the geographical and morphological structure itself becomes a value against which to make a comparison. However, this should be done without any purely archaeological perspective.

Second, to work using the construction of scenarios means accepting that the description of an urban settlement's future state could become a narrative and that fragments of the truth can be contained in this narrative. Both the narrative's evocative capacity and the possibility of exaggerating the meaning of some variables rather than others allow synergy and relationships between different aspects of the structure to be seen which would otherwise be lost or overlooked. This is particularly significant for the territorial structures whose transformation happens in small ways on a daily basis and for which it is fairly easy to lose a sense of meaning, unless it is possible to disentangle the reasoning of the part from that of the whole at the same time. From here is derived the need to reactivate the ability to work towards objectives but, at the same time, towards projects which can examine the physical nature of the urban transformation hypotheses put forward.

In the case of Venice, a further level of difficulty is presented by the fact that the prefiguration of possible and/or sustainable scenarios could not leave some general references out of consideration. First, consideration must be given to a series of large-scale works which concern the very survival of the settlement system, unless catastrophic scenarios are envisaged, and which make up a huge amount of ventures and public works. Sea defences, canal maintenance and the curbing of wave action, the removal of pollution from the lagoon and the regulation of its drainage basin, the construction of the sewerage system in Venice and its completion on the mainland make up the subjects to refer to for any prospective development. Second, the economic and functional reference context on the territorial scale can be nothing but the metropolitan area of the Padua–Treviso–Venice triangle (PaTreVe), within which the proposal for using a sub-lagunar light-railway for Venice has also been made. This question has been a subject of open debate for many years already; a debate fuelled by the hope that the rapid connection between places might become a fundamental criterion in changing the meaning of the different parts' position within the boundaries of the extended territorial structure. Despite the consistent nature of the criterion, the inherent inflexibility of this means of transport and its vocation for serving dense areas with high commuting potential mean that interest in the rationalisation and development of a complex and diversified transport system, which could view the water as an opportunity, takes priority, rather than placing palingenetic hopes on a single variable.

Whilst convention keeps such issues in the background, it is clear that they are by no means irrelevant for the prefiguration of possible evolutionary scenarios for the Venetian settlement system.

4.1. The first evolutionary scenario: the absence of significant decisions

The development of existing trends, in the absence of any decisions, foresees a progressive and apparently contradictory process of becoming uniform and, at the same time, of separation between the different parts, with respect to the broader area of the territory (PaTreVe).

The overall weakness of the system will tend to become more marked and its internal contradictions of being a rigid and composite structure will become extreme. The separation of the four settlement types of Venice, the Estuary, Mestre, Marghera, will increase in functional and political terms. The industrial area of Marghera will continue to maintain its extraneous character of large basic industry, in relation to the production system of the Veneto, the marginalised nature of the estuary will be accentuated and it will be progressively abandoned, whilst parts of the coast-line will look to Jesolo for an increasingly closer link; the Mestre-Marghera system will be subjected to greater tension with the rest of the territory and the neighbouring municipalities. This tension will be translated, on the physical level, into the settlements unravelling along the main routes, thereby prefiguring an undefined conurbation whose central part, Mestre, will (Fig. 6) itself become increasingly inaccessible and perhaps not without degradation. One scenario which foresees, therefore, alongside a general growth of the widespread urban structure and of the progressive uniformity of the urban features without quality and planning, the increase of indifference to location, of land consumption and of the use of private automobile transport.

The counterpart to this progressive loss of identity for the mainland settlement system is the progressive accentuation in Venice's specialisation in which the tourist monoculture will be increasingly intensified, with a progressive marginalisation of all other activities. Without policies, the inflexibility of the Venetian hotel structure will lead to a territorial redistribution of accommodation, just as the mainland city will also be able to accommodate parts of the Venetian university system, thus respecting the theory of creating equally equipped urban fragments.

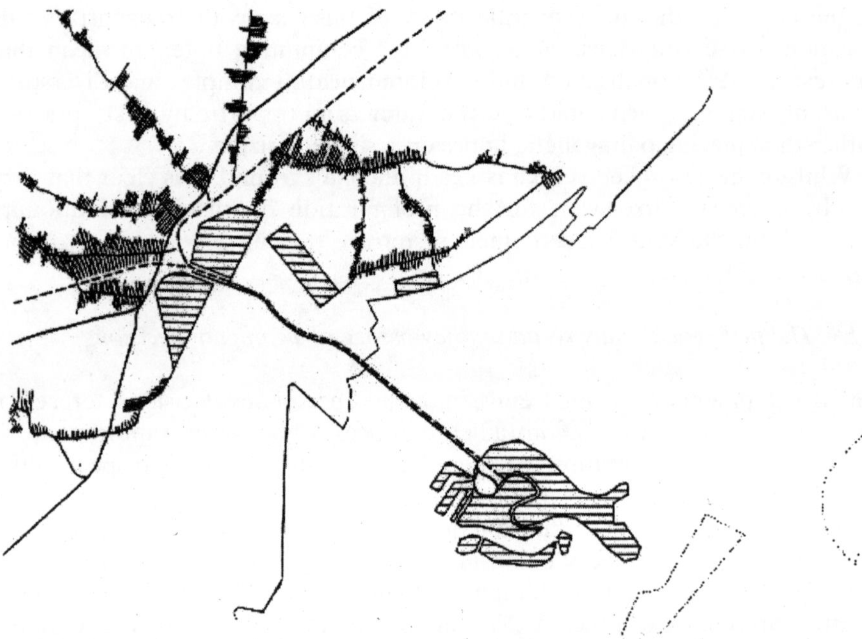

Figure 6 Schematic representation of the evolution of the settlement structure in the absence of significant decisions.

The structural weakness of the demographic pyramid will continue to become more marked, leaving room for second homes or for further abandonment. According to this framework, the idea of a city for the elderly can easily take shape. In this scenario a growth in the non-working population would be obtained, with a possible specialisation in the hospital system. Buildings would increasingly take on the role of undifferentiated boxes and will have to lend themselves to occasionally accommodating the most disparate functions. However much it might be concerned with the physical survival of the city precisely because it constitutes its principal resource, the tourist industry will, nevertheless, contribute only marginally to the process of continuous maintenance, which is usually linked to stable residence, as a result of the works commissioned by residential tourism. This has the effect that the focus of restoration increasingly has an emergency and show-case importance. The transport system will not find reasons for being differentiated in ways which are any different from the present solutions, except for progressive adjustments of the congestion level at the existing bridgehead.

This first scenario of a tourist park is not incompatible with the radicalisation of environmentalist topics in terms of constraints, since the lagoon resource can become the city's natural frame and physically significant changes might generally be prevented. However, the two points of view could, instead, come into conflict where the main issues related to the ways and extent of the use of the resource itself are concerned. In this way, a succession of degradation phenomena caused by an increasingly large exploitation of even the lagoon area for tourism purposes would be set in motion, thus activating a policy based on the imposition of prohibitions and constraints in the effort for damage limitation. The probable outcome of this attitude would, in any case, involve an abandoning with no concern for the lagoon environment and its consequent impoverishment through barely legal activities, including illegal fishing, which use up the lagoon resource since they have no concern for the necessary process of its renewal.

4.2. The second evolutionary scenario: regulated congestion or the game of the parts

In this scenario, the general comparison constituted by the PaTreVe area, would not come about by looking for similarities, so much as by, rather, giving value to the differences and to the environmental peculiarities, thus basing the respective specialisation of the interrelated centres on these. The settlement diversity of the different areas form as reciprocally active resources. Therefore, the objective becomes one of making this diversity synergistically relevant within the creation of a unitary framework of functioning. In short, it is a matter of establishing the unity of a system which is differentiated in its various parts. Central areas are defined as those able to supply the settlements with water and with land at the same time, even if this does not mean that these areas are really central from a geometrical point of view. Deriving from the consideration that the land-water relationship is one of the basic conditions for this centrality is the need to distinguish this condition as a centrality within the system. This might also intervene in the form of curbing the progressive separation-drift in meaning of the parts of the settlement as well as in an external sum of different opportunities for connections on the territorial scale. The internal system is supported by the axis of the bridge across the lagoon and on the two bridgeheads, while the expansion of the whole system finds the principal points of application in the inter-modal (land and water) terminals of Fusina, Tessera and Punta Sabbioni, as well as in the ways of arriving in Venice (Fig. 7).

As a consequence of this system of functioning, the possibility of curbing the external expansion of the settlements by promoting a definite transformation operation, opens up for the mainland urban structures. This operation also allows the possibility of a densification of the central areas and of a renewed transverse connection between Mestre and Marghera at all the points where it is possible for such a connection to become an objective of primary importance. The existing by-pass, supported by the new ring-road, would act as a rapid axis of mobility within the urban system. The *Ponte della Liberta'*, which

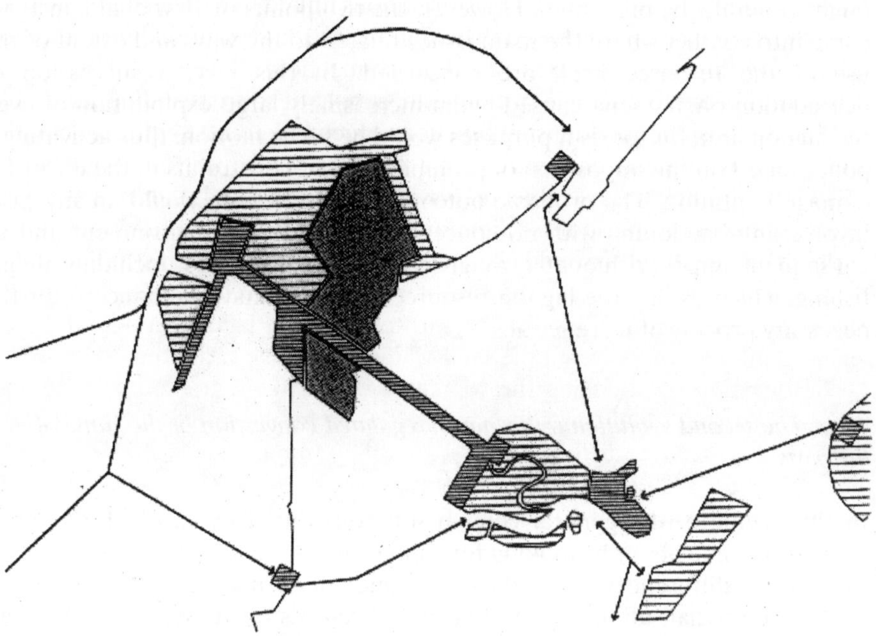

Figure 7 Schematic representation of the evolution of the settlement structure with the hypothesis of 'regulated congestion'.

connects Venice to the mainland, should also be included in this model since the possibility of reaching Venice quickly, whether on public transport or using private transport, must be considered as a fundamental resource for the residents. Setting up three new terminals can improve the management of tourist flows not only through a policy of access regulation but also by means of redesigning the means of arriving in Venice. It is a question of limiting the levels of congestion in St. Mark's Square which are caused by the fact that the majority of tourists find their main fixed points of reference in the square itself, in the Basilica and in the Doge's Palace.

In this way, the re-use of the Arsenal can play a fundamental role in extending the urban system, including the tourist area, towards the littoral and the island areas. Locating the landing of the tourist influxes from the Punta Sabbioni Terminal in the area presently occupied by the ACTV dockyards would lend further support to this scheme. Indeed, 'ever since the Arsenal has taken the form of a mere empty enclosure, it has, in fact, become the cornerstone of a progressive peripheralisation process for the entire district of East Castello. This process is made worse by the central island system's loss of relationships of integration or need with the lagoon and by the decline in the role of the city edges' (Magnani and Messina, 1985). The fact that the Arsenal constitutes a part of the city which is so large and complex that it is not possible to consider it as an isolated structure in any way, makes it possible to examine some feasibility hypotheses for transformations linked to a diversified re-use of its parts which is in no way based on a mono-functional solution but, rather, on the relationship with the axes of urban connections. The size of its spaces makes it possible to envisage not only the supply of tourist landing stages but also the establishment of unusual functions which link tourist-cultural activities (museums) with research centres, also international ones, favoured by the possibility of quick water connections with the airport and, therefore, of residence connected to periodical commuting. By concentrating functions linked to the urban centrality around the trans-lagunar bridge's arrival point and in pushing those with more far-reaching territorial spheres of reference, or for which the time factor is less important, towards the lagoon, a hypothesis of renewed interaction between the land-water system is prefigured.

The question of attacking the weakness of the demographic pyramid can only be entrusted, mainly in the short-term, to public policies. Nonetheless, from this perspective, the city possesses a potential resource in the presence of the university. Therefore, it is a matter of pursuing strategies aimed at the further enlargement of the two Venetian universities, particularly by changing them into residential universities, with the construction of a 'hall of residence' which could have a positive effect on the present situation of competition for housing with the residents. This would lead to the establishment of a young population in Venice, the engendered effects of which are well-known both in terms of daily consumption and for the establishment of services related to games-cultural activities such as sports shows, cinema or theatre, etc., as well as for the establishment of relationships of proximity and of mutual exchange with other research institutions. These can then create connections among themselves concerning the supply and demand of jobs, in a trend towards advanced tertiarisation within the prospective 'Venice, city of learning'.

Freeing a housing policy of theories linked to welfare also means dealing with the very delicate problem of the replacement of buildings which, by seriously altering the essential principles for the continuity of the urban morphological structures, constitute criteria of inflexibility and of contortion of some parts of the building fabric. Extending the view to the lagoon area can, perhaps, go beyond the idea of Venice as a dispersed city, providing incentives for fish-farming which is considered a traditional activity in its own environment. The settlement principle to refer to is that of the point-form settlement separated by large distances which might, however, be able to govern and regulate the territory. It is a matter of identifying economic activities which have their main resource in environmental protection. The tourist industry must be seen as a periodic and point-form complementary activity carried out with full awareness of the fact that the fragility of the environmental system would not sustain mass visiting. In this case, the idea of a park would not be founded on the principle of prohibition and of the consequent abandonment.

4.3. Third evolutionary scenario: the Preliminary Document to the General Modification of the Town Planning Scheme

If the first two scenarios make the respective development possibilities extreme, the scenario prefigured by the Preliminary Document to the General Modification of the Town Planning Scheme cannot but be placed in the gap between the two scenarios. The proposals contained in this document represent a significant effort of elaboration in limited time-periods aimed at ordering the confused stratification of Venetian urban-planning and at promoting renewed considerations of the city. The mechanisms for starting it are faced with four difficulties of different levels.

First, the need to operate in limited time has not permitted the establishment of a systematic investigation which could set up innovative procedures starting from the analytical level. In fact, it was rationally pursued by utilising the enormous existing volume of descriptive materials on Venetian phenomena. These are the products of existing Venetian research bodies or drawn up at the time of previous technical Modifications, which, however, have not always been produced within a systematic framework of investigation which is conducive to the construction of planning instruments.

Second, the need to deal with a multiplicity of town-planning instruments in force has had the effect that it has proceeded at different degrees of progress according to the most evident deficiencies within the composite framework of the existing plans or of the various emergencies in the planning framework in force. In fact, dealing with a unitary planning of the territory is attempted for the first time since the approval of the 1962 Town Planning Scheme. Territorial planning until today has, instead, proceeded by means of a rapid succession of partial modifications (which often have not even completed their formal procedure) and measures of varying degrees of urgency which are placed on the territory in the absence of a unitary idea.

Third, the hypotheses of the Preliminary Document represent the intention, expressed in the planning report, to give concrete planning direction to the framework of ideas and suggestions which emerged in the course of research conducted by the Istituto Gramsci at the end of the 1980s, and were publicly discussed at the two conferences in 1988 and 1990 (Var, 1990, 1988).

Finally, the hypotheses contained in the Preliminary Document of the Scheme can, after due consideration, be deemed a scenario in as much as they are a complete prefiguration of a territorial structure. Nevertheless, in contrast to the preceding scenarios, its capacities for being valid lie within the limits of the town-planning itself. Indeed, the plan does not act directly on all of the mechanisms at work in the urban area, so much as, rather, on the conditions which allow the development and regulation of the phenomena. Therefore, the planning documents are a mixture of aims and actions which are possible to implement. The elements used to put the aims into effect are made up of planning techniques which concern the regulation of the types and of the degree of land use. Only in some cases do such techniques indicate precise actions and, in any case, the difficulty of establishing implementation priorities remains.

Whilst this is not the place for constructing a system of judgement or assessment of the planning documents it is possible, however, to try to reread the bones of the choices contained in them in light of the considerations developed up until now, in particular with the use of the keys proposed above. Within the current debate it is rather unlikely to be able to draw theoretical basic principles which inspire a formally completed idea of the city to which a single and unitary planning technique might correspond, consequently, as a tool for its realisation. As has already been confirmed, the existing city and its (internal) transformation constitute, so to speak, the whole field of application of the consideration.

It can be stated that the hypotheses formulated from the planning documents bear witness to this work prospect, by finding the inspiring criteria for reference in the idea of sustainability which is understood as the identification of and giving value to the existing resources within a framework of synergetic and systematic optimisation of investments. In this sense, the idea of sustainability, with its inherent attention to the environment, can penetrate all planning choices without having to be limited to marginal considerations or additional ones. Really, the attention given to problems of an environmental nature does not become a specific intervention sector, without understimating the pertinent implementation techniques, but assumes the form of a general criterion which is capable of creating a constantly present attitude.

All this allows us to envisage a framework of changes which could utilise a multiplicity of techniques for intervention which are appropriate for every specific problem and yet still coherent with the realisation of general objectives. It is within this framework that the frequent references to the functioning of the public administration contained in the planning document must be read, on the one hand, as an exploration into the limits of town-planning's potential as a planning technique, and, on the other, as a call for the creation of total consistency in the administrative acts for reaching the general objectives.

From these kind of considerations the formulation of a comprehensive model for territorial structure transformation is obtained which is based on the strengthening of the framework for the system's unitary functioning. The objective to construct the centre of the bipolar city is the fundamental expression of this idea. This entails the radical transformation of the areas connected to the two bridgeheads of land and water joined by the axis of the *Ponte della Libertà* and the railway bridge. The T-shaped model for functioning, which stretches as far as including the areas of Via Fratelli Bandiera in Marghera, proposes an internal urban system which has the objective of removing a series of large, existing, inflexible elements constituted by the presence of an internal periphery, by the absence of suitable connections between Mestre and Marghera and by the prevention of the mainland city from facing towards the water. This proposal is completed by the project for the establishment of terminals at Fusina, Tessera and Punta Sabbioni which complete the model for general functioning.

In the historical city this causes a shift in functions towards the bridgehead, while for the mainland it is translated into an increase in the intensity of use of the different parts with greater potential for connections with the entire functional model. From this ensues a hypothesis for the containment of the urban area. This is brought about by means of a policy for making building more dense in the central areas of connection and, consequently, of limiting and consolidating expansion towards those areas of the territory which are not yet urbanised.

Potential factors of inflexibility can be found in those proposals in which the impromptu research into the availability of areas seems to predominate, such as in the case of the location of the football stadium at Tessera, of the university buildings in Via Torino in Mestre and, perhaps, of the Mestre hospital, more than on the development of synergy with the whole settlement system for which, furthermore, the objective to establish transverse links between Mestre and Marghera does not seem to be pursued with the required determination. However, the inflexibility of the settlement structures can also manifest itself as a consequence of the rejection of opening up opportunities. In this sense, the idea of a conclusion of the Venetian building potential leads to the rejection of the transformation of portions of the building structure, even within the framework of a more defined housing policy. It also leads to the proposal to create green areas by making reference to problems of urban-planning standards, without, however, considering the peculiar nature of Venice, city-park, as a whole. This attitude is seen particularly in the case of the decommissioning of the port areas of St. Basilio and St. Marta. Whilst for the problem of student accommodation, we stop at the immediate opportunities without considering them within the more comprehensive framework of Venetian accommodation. It is enough to think that measures are envisaged for creating 300 places to stay for a situation which in 1993/4 saw about 1100 apartments occupied by students (Bonifacio *et al.*, 1995) and which today sees an increase in this figure to 1657 private dwelling units occupied (Dina, 1996).

What is much more significant is the suspension of the decision on the Arsenal which makes it possible to avoid a policy for cultural establishments in conjunction with the specification of the methods for landing the tourist flows coming from the terminals, as if these in themselves could resolve the congestion problems in some parts of the historical city. This is particularly precarious because, without this part of the decisions, the moving of such important functions towards the bridgehead appears as an inert element of the city's progressive emptying, rather than as the result of a plan which is able to interpret the vocations of the different urban parts within a unitary framework.

At the present state of elaboration, in some cases the planning proposals only make investments in terms of urban design. This is the case with the issues of curbing urbanisation and of redefining the urban borders with agricultural territory, as well as that of making the central areas of Mestre more dense.

Even the idea of the 'Mestre wood' project could be read in terms of territorial planning if it were able to establish a capacity for morphological dialogue with the urban settlement structures with which it comes into contact on the one hand, and, on the other, if it could open itself up towards the structures of a territorial nature, ranging from the agricultural structure to the geographical-hydraulic one, within the scope of a comprehensive programme for upgrading the environment.

Bibliography

Alberti, M., Solera, G. and Tsetsi, V. (1994) *La Città Sostenibile: Analisi, Scenari e Proposte per un'Ecologia Urbana in Europa*. Milan: Franco Angeli.

Albrecht, B. and Benevolo, L. (1994) *I confini del paesaggio umano*. Rome-Bari: Laterza.

Augé, M. (1993) *Non Luoghi*. Milan: Elèuthera.

Benevolo, L. (editor) (1996) *Venezia. Il nuovo piano urbanistico*. Rome-Bari: Laterza.

Bettini, V. (1986) *Elementi di Analisi Ambientale per Urbanisti*. Milan: CLUP.

Bettini, V. (1995) *L'Impatto Aambientale Tecniche e Metodi*. Naples: CUEN.

Bettini, V. (1996) *Elementi di Ecologia Urbana*. Turin: Einaudi.

Bettini, V., Falqui, E. and Alberti, M. (1990) *Il Bilancio di Impatto Ambientale*. Milan: CLUP.

Bonifacio, M., Vassallo, M. and Zanus Fortes, L. (1995) *Indagine sulla condizione abitativa a Venezia degli Studenti Fuori Sede*. Venezia.

Campos Venuti, G. (1995a) Il preliminare del PRG di Reggio Emilia. Le innovazioni del piano: perequazione ed ecologia. *Urbanistica*, **103**, 66–75.

Campos Venuti, G. (1995b) L'ecopiano di Reggio Emilia. *Verde Ambiente*, **V**, 24–29.

Consorzio Venezia Nuova. Information Service. (1995) *La Laguna*, annexed document of the preliminary plan for the new 'Piano Regolatore Generale' of the municipality of Venice, Venice.

Dina, A. (1996) La politica della casa. In: Benevolo, L. (Ed.) *Venezia. Il nuovo piano urbanistico*. Bari: Laterza, 1996

Istituto Ricerche Ambiente Italia. (editor) (1995) *Ambiente Italia 1995*, Milan: Ambiente Editions.

Gregotti, V. (1986–1993) Il valore politico del disegno urbano. In: Gregotti, V. (Ed.), *Le scarpe di Van Gogh*. Torino: Einaudi.

Gregotti, V. (1994) Il valore politico del disegno urbano. In: Gregotti, V. (Ed.), *Le scarpe di Van Gogh*. Turin: Einaudi.

Karrer, F. (1993) *Progetto di strade: Una rassegna di esperienze*. Cosenza: Pellegrini Editore.

Lawrence, G. (1996) Il caso studio dello sviluppo del Seattle General Plan. Presented at the Seminar *La Sostenibilità del General Plan di Seattle*, DAEST-IUAV, Venezia, Ca' Tron 14/18 October.

Linch, K. (1981) Le dimensioni prestazionali. In: Linch K. (Ed.), *Progettare la Città. La Qualità della Forma Urbana*. Milan: ETAS Libri.

Lucan, J. (1993) Projets urbains pour les grands ensembles. *AMC*, **41**, 19–30.

Magnani, C. and Messina, F. (1985) Note sul rapporto Arsenale città. In: Gennaro, P. and Testi, G. (Eds.), *Progetto Arsenale*. Venezia: CLUVA.

Migliorini, F. (1989) *Verde Urbano*. Milan: Franco Angeli.

Munda, G. (1995) La sostenibilità socio ambientale dei sistemi economici. *Urbanistica*, **104**, 40–49.

'Rassegna' (1985) 22, *Venezia città del moderno*, monograph.

Romanelli, G. (1988) *Venezia Ottocento*. Venice: Albrizzi Editore.

Romanelli, G. and Rossi, G. (1977) *MESTRE. Storia, territorio, struttura della terraferma veneziana*. Verona: Arsenale Cooperativa.

Rosa Salva, P. (1974) Trasformazioni ambientali ed alterazioni nella Laguna Veneta. *Urbanistica*, **62**, 6–44.

Salzano, E. (Ed.) (1992) *La Città Sostenibile*. Rome: Edizione delle Autonomie.

Secchi, B. (1989) Grandi spazi aperti, In: *Un Progetto per l'Urbanistica*. Turin: Einaudi.

Secchi, B. (1994) *Il racconto urbanistico*. Torino: Einaudi.

Secchi, B. (1995). *Urbanistica Quaderni, 4*, monograph on landscape.

Secchi, B. (1996a) Una teoria per l'ambiente. In: *Laboratorio Prato PRG*. Florence: Alinea.

Secchi, B. (1996b) *Una Progetto per Prato*. Florence: Alinea.

Urbanistica (1968) monograph on Venice, 52.
Urbanistica (1972) monograph on Venice, 59–60.
Urbanistica (1990) 98, Franco Angeli Milan.
Urbanistica (1990) 100, Franco Angeli, Milan.
Urbanistica Quaderni, (1996) 9. *La Pianificazione Urbanistica come Strumento di Politica Industriale. La Variante al PRG per Porto Marghera*.
Var. (1985) Studio di fattibilità del progetto di recupero dell'arsenale di Venezia. In: Gennaro, P. and Testi, G. (Eds.), *Progetto Arsenale*, Venice: CLUVA.
Var. (1986) *Quattro Progetti per l'Area Ovest*. Venice: Marsilio Editore.
Var. (1988) *Gli Indicatori Ambientali. Valori, Metri e Strumenti Nello Studio di Impatto Ambientale*. Conference proceedings 1984, Milan: Franco Angeli.
Var. (1988) Idea di Venezia, In: *Quaderni della Fondazione Istituto Gramsci Veneto, Conference Proceedings 1988*. Venice: Arsenale Editrice.
Var. (1990) La nuova dimensione urbana: Venezia-Mestre nella regione Veneto. In: *Quaderni della Fondazione Istituto Gramsci Veneto, 9: conference proceedings 1990*, Venice: Arsenale Editrice.
Var. (1993) *Costruire a Venezia*. Venice: Il Cardo Editore.
Var. (1996) Case vuote, case inutilizzate, seconde case: i risultati di un'analisi delle utenze ASPIV, ENEL, Venezianagas. *Osservatorio Casa*, September.

CHAPTER 5

Material Production in the Municipality of Venice

Maurizio Rispoli, Francesco di Cesare and Andrea Stocchetti

1. Some aspects of the historical evolution of the presence of material production in Venice

For convention's sake, we have defined material production not only as production activities, the output of which we usually call goods (primary materials, intermediate products, parts, components, investment goods and consumer goods, both durable and non-durable) as opposed to services, but also all those activities which, in any case, are characterised by a specific tendency to be land consuming, such as wholesale trade and, in particular, the coastal fuel deposits, port and airport services.

Despite the fact that immaterial production has significantly increased in qualitative and quantitative importance over the last two decades on a global level, an initial question must be asked; whether the production base of an urban area can be substantially characterised and sustained in the future, from the socio-economic perspective, only by other non-material production activities. An initial response, which remains to be proved, could lead to the belief that, despite the fact that material production has continually declined over the years in all of the urban areas of the oldest industrialisation, such as the Venetian area, it still seems possible and, perhaps, beneficial to the socio-economic and cultural equilibrium of the city that material production not only contributes to generating the demand for labour and, therefore, employment, but also contributes to differentiating the export base with respect to tourism-oriented supply and other possible immaterial production activities which are characterising many metropolitan areas.

We now turn our attention to the Venetian area. The data we have, which relates to the situation of the production plants at the end of 1991 (ISTAT census) and the information gathered from different sources referring to more recent years, allows us to formulate an initial descriptive-interpretative picture

I. Musu (ed.), Sustainable Venice, 121–157.

of the importance and the role of material production within the Venetian social, economic and environmental system.

As a whole, the data for the beginning of the 1990s shows that material production in Venice still represents a considerable portion of the production structure of the area studied, in terms of employees but also certainly in terms of value produced (31.8% of employees and 31.3% of local units). The greater part of this production structure is concentrated in Porto Marghera. However, the importance of this industrial area for material production in the Venetian territory must not make us lose sight of other activities which are of relative importance for the city and, in particular, for the lagoon areas (the historical centre and the islands).

Some of these production activities, which generally accompany the economic and social organisation of any urban conglomerate of some importance, supply consumer goods for the population and on the whole have the features of artisan enterprises (bakers, pasta makers, pastry makers, general food production companies, building maintenance and construction companies, blacksmiths, carpentry workshops, furniture suppliers, etc.) and do not have any specific features. As a result, these can be ignored for the moment.

Others are distinguished precisely because they are present in the Venice municipality for reasons related to specific location factors and are, in general, not found in other cities of the Veneto. We refer above all to:

- the various major and minor shipyards present in different parts of the city other than in Porto Marghera (the Arsenal, Pellestrina, Giudecca, St. Elena, Lido, Burano, Murano, etc.) which are strictly related to the waterfront;
- the aircraft workshops which are involved in the transformation and extraordinary maintenance of aeroplanes and can only operate in areas which are adjacent to or in direct communication with a large international airport, as in the Venetian case;
- the fish-farming and valley fish-farming, typical of the lagoon environment.

Finally, there are also other material production activities which are situated in the Venetian area as a result of the influence of aggregated historical location factors. This is the case, for example, for the furniture manufacturers and the gilders in the historical centre and for lace production in Burano as well as Murano glass production which stands out for its particular production concentration. In fact, on this island, we find the highest index of specific industrialisation (employees in a specific production sector/resident population), which is glass making in this instance, but also the highest index of industrialisation for material production (0.42) out of all the districts, including residential ones, into which we have divided the territory (average for the districts is 0.08).

As a whole, all of the material production activities mentioned above must be considered as distinctive and characteristic elements of the area and, within the limits set by the principles of efficiency and effectiveness together, must not be considered producers of goods and income alone, but also of culture.

Material production in Venice is not a recent development of the twentieth century, but also largely characterised the latter part of the last century and the early years of the 1900s. As has already been noted, material production in Venice certainly did not arise with the creation of the industrial area of Porto Marghera, but exists before it and specifically concerns the lagoon and island part of the present municipality of Venice.

After the fall of the Republic, from the 1830s onwards industrial development took place. Such development was driven on the one hand, by the desire to bring about the rebirth of a commercial port in Venice and, on the other, provided the chance to be part of the intense economic-industrial progress which characterises Europe from the second half of the last century.

After the brief period of French occupation in Venice, during which there were rather important town-planning works (the Napoleonic Gardens, the beginning of the Via Eugenia, then Garibaldi, the Napoleonic wing of St. Mark's square, etc.), 50 years of Austrian domination follow (1815–1866). Although there was no incentive for taking port operations up again during these years since Vienna's policy favoured the ports of Trieste and Fiume, the isolation of Venice from the mainland was terminated by the construction of the railway bridge (1834). Within certain limits, this connection made this different and unique city the same as other continental ones and, in terms of rail transport, is the symbolic triumph of the technological progress of the nineteenth century.

In the Austrian period, during which an overly cautious innovative policy caused a stagnation in the city's economic and social evolution (the population remains around 110,000 over this period), there were, nevertheless, other isolated works of some importance. These include the construction of the two iron bridges (Scalzi and Accademia) in addition to the existing Rialto Bridge in order to facilitate crossing the Grand Canal (1854–60), the start of a sufficiently wide and straight route in the Cannaregio district, starting at the railway station and leading to the heart of the city as well as the construction of the General Slaughterhouse at St. Giobbe (1842). This single site unified slaughtering activities, which were dispersed in different places in the city until the middle of the 1840s.

However, after unification with the Italian State, in the period until just before the First World War, new production activities for goods and services were set up and the pre-existing activities start again, such as the small and medium shipyards, production of tobacco and cigarettes, glass manufacturing, the tanning industry. There was, however, a feature, above all of the new initiatives, that should be highlighted: these are mainly the work of entrepreneurs and capital which are extraneous to the old entrepreneurial and commercial classes of the Republic and, furthermore, are often not Italian, as the names of many enterprises often show.

The new initiatives' activities, as well as those of some of the pre-existing ones which undergo expansion (tobacco), can in some cases be seen to be directly linked to the opportunities opened up by the planning and construction of the Stazione Marittima (1880) which allows a rapid re-conquest of the port operations that the city lost during the preceding decades (Dorigo, 1973, p. 58).

Among the different material production activities preceding the birth and development of Porto Marghera, it is useful to remember those which created plants of dimensions which can be considered medium-large in terms of employees by the standards of the era.[1] The E.G. Neville foundry and mechanical construction plant, located between St. Rocco and the Tolentini, produced machines, heaters, metallic structures and structures for bridges which were both useful for other enterprises in the city as well for urban projects. This employed about 400 workers in 1887[2] but expanded further in subsequent years. The Manifattura Tabacchi, which existed in its current location before the fall of the Republic, expanded greatly in the decades referred to so that, still in 1887, it employed about 1740 people who were mainly women. The S.A. Cotonificio Veneziano developed in an area of St. Marta which is easily accessible for supply and deliveries. Production began in 1883 and some years later, therefore with reference to 1887 again, the number of employees was already higher than 900. However, it expanded further at the beginning of the 1900s.

In the textile sector again, the Jesurum company was really large in terms of employees (about 2500 people, the majority of whom were women, again around 1887), but was scarcely mechanised. It was located at the edge of the Castello district near St. Mark's.

The match factory, Baschiera, began production in 1875 in the St. Giobbe area, which then led it to be one of the most important in Italy within a few years and to provide work for about 750 people in 1887. This number subsequently grew, particularly at the beginning of the twentieth century, after the transformation of the private firm into a company called 'Saffa' and after the reconstruction and expansion of the old plant. The grain-milling plant located at St. Biagio on Giudecca, which Giovanni Stucky set up in 1883, already had 187 workers in 1887. He subsequently expanded it several times, adding further buildings until the first decade of the 1900s when the structure which is still recognisable today was obtained. The watch factory which developed next to the *Ponte Longo* on Giudecca in 1878 and assumed the name of 'Arturo Junghans and Hans Herion' in 1899. It was founded by the Herion brothers who were general agents for the entire Italian market of the Gebrueder Junghans di Scharnburg in Wuertemberg. The Vittorio Narduzzi brick and lime factory, operative from 1874 which was, for more than half a century, the most important factory for building materials in the city (it possibly employed about 200 workers, still in 1887). The Ceresa-Millin firm in the St. Leonardo area in Cannaregio which, having expanded the previous Zecchin and Ceresa plant, worked for a number of years in the tanning process and in the grinding and refining of sulphurs and employed about 150 workers in 1887.

Finally, plants which were generally mechanical workshops, foundries and shipyards at the same time are taken into consideration. Above all, we refer to the Arsenal which, located in the Castello district, employed about 3851 workers in 1887. In addition, the F. Layet mechanical plant with foundry and shipyard, located in *Campazzo delle Erbe* where the Riva dei Martiri is today. Here, various machines, turbines, vapour heaters and boats of medium tonnage were produced. About 60 workers were employed in 1887 and this number grew over the following decades. (In 1915, the shipyard SVAN with four production floors arose adjacent to this area. This specialised in the construction of torpedoes and had a brief life due to both the end of the war and the area's urban restructuring). Finally, the *Società Veneta per Imprese e Costruzione Pubbliche* must be mentioned. From 1872 onwards, this company operated for about three decades at St. Elena in an plant which included a shipyard and a mechanical workshop which employed, again in 1887, about 850 workers.

Altogether, the medium and large material production activities just mentioned provided work for about 10,000 workers in the last decade of the nineteenth century. The employees of the many small and artisan businesses as well as those from the glass industry, mainly concentrated in Murano, have not been included in the plant considered previously and must, therefore, be added to this figure. The production activities which have been mentioned were mainly located in the western area of the city, in an arc that goes from West Giudecca to St. Marta, Campo di Marte to St. Giobbe. This settlement pattern seems, in fact, to be influenced by the effects of two attractions which linked the city with the rest of the world: on the one side, the commercial port with the Stazione Marittima, on the other, the railway station with the trans-lagoon bridge.

The start of a mass of great technological innovations and the introduction of new organisational solutions into the production process (Taylorism and Fordism) which have characterised industry since the last few years of the nineteenth century, combined with the appearance of completely new industrial sectors, was unable to find applications and location solutions in Venice's historical centre. Over the years, the existing production activities lost competitiveness in relation to economies of scale, to continuous processes and to the possibilities opened up by road transport. As is well known, the evolution of material production in Venice, in fact, opted for the mainland where the above-mentioned advantages can be easily obtained. Furthermore, to these advantages must be added the brilliant idea which led to the conception of unloading ships as the first stage in the production processes, working autonomously, in the plants which grew from the 1920s onwards in Porto Marghera.[3]

2. Situations and scenarios

To envisage the more probable scenarios for the evolution of any complex phenomenon is not an easy task. This is certainly the case where the subject of the investigation is a production system such as the Venetian one within which strong connections between economic, social, natural, hydro-geological, financial, legislative and political aspects develop. This complexity appears particularly striking in such a context since it is a result of the numerous variables at stake, some endogenous, others exogenous to the processes of choice which the multiple public and private decision-makers are involved in.

It is particularly complicated to tackle the subject on a scale which, although restricted, sees the co-existence of many different situations; since these are characterised by history, constraints and particular opportunities which need to be treated, at least in part, separately. In spite of this, the problem of equilibrium, which is implicit in the very idea of sustainable development, requires considering the area as a whole. Some exceptions occur for those specific topics which characterise the single zones which make up the municipal area. These areas can be identified as: island Venice's historical city; Venice lagoon; mainland Venice's urban centre and its suburbs; seaside resort and coastal Venice; port and industrial Venice.

The attempt to outline the evolutionary scenarios deemed probable can only begin with an analysis of current and past tendencies which aims to distinguish between conjecture and trend and gather the signs of change which can transpire from some phenomena. The reading and processing of the data available has made it possible to grasp some important aspects which are now considered quickly and will be subsequently highlighted in this chapter.

In terms of employment, material production's importance out of all production activities is much lower than the regional average, but its value, in terms of employees, which is greater than 30%, is sufficient to disprove the theory of those who would see this sector as a shadow of its former self and, therefore, negligible today. In fact, the city's morphological features and its strong vocation for tourism are enough to explain this peculiarity and, as a consequence, the difficulty in interpreting the data on production in relation to the traditional reference parameters, such as, for example, the regional averages. These, as it is well known, reflect very different contexts.

In fact, in observing the structure of material production in the Venice municipality and comparing it with the Veneto, its atypical nature emerges which can be summed up as the fundamental absence of that which is defined as the Northeast model. This is a system which is sustained by the existence of numerous small-medium enterprises which make their specialisation and interrelations between processes and chains the basis of their success. As far as employment is concerned, in the municipality of Venice the leading role is given to a few large enterprises which have been and still are important parts of the city's production history.

Despite the general crisis (in the whole industrialised world) in the same basic and process industry which has characterised the development of Porto Marghera in this century, there are some weak signs of upkeep and of revitalisation. Indeed, there are some which depend on a radically innovative approach for the sector to which they belong and others which are simply able to maintain a dynamic equilibrium with their own reference environment.

In terms of the quantity of local units, it is, rather, the existence of small enterprises, above all artisan ones, which appears significant (22.7% of the local units have only one employee and about 90% have less than ten employees). This aspect takes on a particular significance when considering the future of the sector which is not bound exclusively to the industrial area of Porto Marghera. The scarce but significant data regarding this is consistent with this picture. The primary factor for success generally appears to be the combination of a striking entrepreneurial quality with the involvement of many stakeholders in a plan for revitalisation. The diversity in terms of average size, together with a striking production specialisation, contributes to identifying a common geography for the activities and for the plants. This is particularly important for identifying the real possibilities and advantages of development, or of simple maintenance, for some production structures.

Finally, among the relevant aspects, the allocation of material production according to its destinations of use must be pointed out. This highlights the capacity, also for export, of the Venetian production system: 75% can be attributed to goods destined for production and the remaining 25% to goods destined for consumption. The considerable difference can be interpreted in support of the theory which sees location factors as the principal determining factor in choosing to locate in Porto Marghera, also by virtue of the interrelations between the activities present there. On the contrary, the percentage of goods for consumption can also be explained as being, to a large degree, the result of a densely populated urban area.

Other important aspects contribute to defining the current state of material production in Venice:

- the fall in employment, which varies according to the areas and sectors considered, but is taking place anyway in all of the areas with large industries and a high incidence of basic production;
- the trends of the production plants in terms of creation and closure of the enterprises which also varies in relation to areas and sectors;
- the urban and environmental restrictions which, in fact, have conditioned and continue to condition the location and relocation processes;
- the initiatives supporting entrepreneurial activities which aim to encourage new plants and the conversion of plants and of entire areas.

The tourism monoculture tends to be one of the subjects which dominate the discussions on the future economic-production order of the city. However, it is true that the statistical surveys (although if in reality, a careful look at the life of the city would be enough) leads one to consider this shift as being far from impossible. Against this danger of economic and social imbalance, the search for the correct mix between specificity and production propensities appears to be the only way to guarantee the city a development which works for the welfare and the quality of life of the users of the city itself, and foremost amongst these, the residents.

With regard to the specific trends in the production structure which have taken place over the last few years, individual sectors and individual areas of material production structures must be distinguished in the municipality of Venice. Although the industrial area of Porto Marghera undoubtedly constitutes the most important industrial nucleus, which will be justified in an appropriate paragraph, the activity of the Venice Port is also closely related to the fate of this area. However, the numerous and important production activities which are carried out in the historical city, in the islands and on the mainland must not be forgotten. Therefore, we will proceed by dealing with the problems of these areas separately, and then add a general account of the most important production sectors distributed in the municipality of Venice, without location distinctions.

However, it must be said that, while for Porto Marghera it is possible to make a comparison based on recent data which essentially corresponds with that of the census survey in 1991 since a database specifically dedicated to this area exists (Pugliese, 1996), for the rest of the municipal territory, only a comparison based on heterogeneous data from 1991 onwards is possible. Therefore, this must be evaluated with the necessary caution.

With regard to the Municipality as a whole, the impression obtained from a comparison between the census data and the analysis made on the Chamber of Commerce Company Register is that the reduction in the number of employees in the different material production sectors takes place practically across the board, with the noteworthy exception of some land consuming commercial activities (vehicles and fuels trade: +26.2% employees between 1991 and 1996), of construction (+23.3%) and in the metal products sector (+21.5%).

3. The Porto Marghera industrial area

In this setting, the undeniable fact is that Porto Marghera's production structure has declined further since October 1991 (the census date) and it has been ascertained that material production in that area has gone from employing of 17,718 workers when surveyed by Istat (October 1991) to 13,911 on 31.12.94, while today's total (31.12.1996) is estimated as about 13,500 employed.

Important changes have in fact contributed to this, such as the definitive closure of the primary aluminium plants (ex-Sava and ex-Alumix), sited in the first zone of Porto Marghera and which characterised the industrial area from the 1930s. In addition, the closure of the plants for the production of anodes for electric ovens from the Alusuisse primary aluminium plant, the shut-down of the plants for nitrate-based fertilisers production (Agrimont), for coal refining (Italian coke), for iron and steel production (Ilva), for part of the production of heavy metal framing (ex-Cipi-Sartori and Metallotecnica).

In the present circumstances, if the necessary and enormous investments were made, particularly in the petrochemical industry, Porto Marghera's production structure would perhaps find an equilibrium in terms of employment, settling at a figure which is probably lower than that estimated on 31.12.1996, but not below 90% of this.

By just looking at the employment situation, we could say that the industrial area has been declining, from every point of view, to one-third of its maximum, which was operative during the first half of the 1970s (before the oil crisis). The reality appears less uniform and unidirectional than it seems to be at first sight and the extent of de-industrialisation must be interpreted in the light of two interrelating variables. On the one hand, technological progress which has affected all sectors and, therefore, also those which characterise the typical coastal industrial areas developed until the 1970s. On the other, decisions to convert or to relocate made by the large enterprises operating in the basic sector (process industries).

Although the technological progress of the last two decades, combined with the changes in the composition of the production mix of the area, has brought about a drastic reduction in employment, at the same time, it has significantly increased the value added per worker and the earnings per worker. In fact, the highly evident process of Porto Marghera's de-industrialisation has, nevertheless, been less intense than it is often believed and maintained, since it is not correct to measure it only on the basis of the number of jobs. In the whole industrialised world, large enterprises, in particular those operating in process industries, have seen an overall reduction in employment: but what has happened to the industrial production indices? Often, these are not even lower, sometimes they are actually higher, in others cases they are lower but with negative percentage variations which are much lower in relation to the employment trends.

Today, the industrial area of Marghera evidently appears very different to that of about 25 years ago, both in its production composition as well as in the enterprises which are the leaders of the plants present: less state participation, fewer initial phases of the production chain (basic metallurgy, for example).

What still characterises the area and makes it different from any other in the Veneto is, on the one hand, the development of the area in contact with the waterways (the docks) and the spaces which are still available for production plants or have become so over the years; on the other hand, some production activities which are rarely present or completely absent in the rest of the region still take place in the area. These include oil refining, the petrochemical industry, synthetic fibre production, intermediate aluminium products (the products of the processes of lamination, extrusion and drawing),, the shipbuilding industry with the production of large ships including passenger ones.

There have recently been some indications of a turning point which, however, is difficult to interpret until some problems connected to the protection of the lagoon environment are made clear and solved.

The end of year report for 1995 produced by the *Ente Zona di Porto Marghera* (Porto Marghera Area Body) shows the strengthening of some positive indications, which include:

- a modest increase in employment after years of continuous decline;
- a modest increase in goods handled;
- the interest shown in the area by some large foreign groups;
- the activities of Promomarghera;
- the creation of the Science and Technology Park.

At the same time, this report emphasises what appear to be the more important negative aspects:

- infrastructure, with particular reference to the practicality of the port and the links with the road network and railway;
- tariffs for permits and public services which are still too high.

For the operators, some big issues have determined or can, in fact, determine the role and the future of the district. These include:

- the intermodal/interchange system;
- Venice's international role;
- the possibility of the area's conversion.

where the first identifies the principal location advantages of the centre, the second singles out the possibility of attracting attention and financing at the international level and the third, by focusing attention on the crisis that has, perhaps, by now ended and on the most debated subjects, foresees the possibility of redesigning the order of the area.

In terms of intermodality of transport, some progress has been made in matters of air connections, also as a result of collaboration between the airports of Venice and Treviso and the plan for a terminal. Today, Venice can rely on a well-served airport with an intercontinental link 'mediated' by European flights. The problem of links with the motorway network is more complex. Moreover, the function of the freight yard is still to be developed while the network of waterways has not yet been completed.

On the transport front, the role of the port appears crucial as do the real possibilities for exploiting it effectively (a specific paragraph will be dedicated to this subject).

Naturally, Venice's international role has, in one way, borne its fruits through the *Legge Speciale per Venezia* (Special Law for Venice) and the inclusion of the area of Porto Marghera into Objective 2 of the European Community. However, the obstacles which until today have held up any revitalisation of the centre are of another kind and, as is also said below, have in fact impeded the start of a real conversion of activities.

Today, the whole area is about 2000 ha, divided into about 1400 ha of industrial activities (zones I and II), 343 ha of canals and water bodies, 120 ha of commercial port, 77 ha of roads, railways and services and 38 ha for the state-owned belt. The composition of the existing activities form a particularly complex centre due to the co-existence of basic production, research activities, services for businesses and the population, enterprises inserted in production chains and linked to other systems external to the area and the rather heterogeneous dimensions. Furthermore, a distinguishing feature is the presence of heavily land consuming enterprises and, at the same time, the presence of empty areas or those occupied by abandoned plants.

The numerous Protocols of Intent signed by local government authorities and the other bodies officially involved in the management of the area bear witness to the interest in Porto Marghera. Plant closure, transfer of activity or failure of some enterprises all seem to be essentially over, while a new geography of the area has appeared where almost all of the enterprises still operating there should be considered economically healthy, capable of competing in the market and making profits, thanks to the choices made in terms of the product-market relationship and technology. Some sectors act as leaders and represent the basis for the centre's role, not only historically, but also for its future. These include ship-building, oil (refineries and deposits), petrochemicals, the intermediate aluminium products and electricity production.

As is true for other areas of the Venice Municipality, there are also some issues which seem vital for Porto Marghera. These include the system of transport/access and the strengthening of the terminals. The environmental issue has always played a prominent role: currently, after decades of scarce vigilance and dubious behaviour on the part of those responsible for some of the plants, continuous monitoring seems to confirm that the terms of the law are respected, thereby indicating that Porto Marghera is one of the safest industrial centres. Actually the uncertainty comes from the probable future modifications of threshold values: existing industries may well find themselves unable to respect the new parameters.

In terms of the revitalisation of the area and of the new plants, as long as the conditions dictated by the regulatory plan remain unchanged, focus is placed on the possibility of attracting new investors and forming an incubator for small and medium enterprises to work alongside the large ones which are already present or should want to locate in the area[4]. The fundamental issues which have prevented the arrival of new enterprises are the high costs demanded for the areas and the uncertainty about the future of the centre. The choice to locate in Porto Marghera is obviously the result of the opportunity to make use of the location advantages derived from the infrastructure system, from the possibility to face the sea and from the economies typical of an industrial district. However, when confronted with the excessive costs of the areas and, above all, with a situation of uncertainty about the public policies and future control parameters, businessmen tend to look for alternative solutions. Moreover, the areas are subject to inspection of conditions and remediation work is also obligatory. In this case as well, the costs and, above all, the lack of clarity about the methods for soil analysis contribute to discouraging private initiatives.

However, the subject of a lack of clarity should be applied to the whole decision-making system which will inevitably influence the future of Porto Marghera. In the absence of a clear definition of the context's conditions, including the future nature of sea and land access, and without a development plan for the area, businesses are not in a position to make strategic decisions concerning the location of their activities.

It is on this issue, that the design of a future scenario for Porto Marghera depends, along with other equally important ones, such as the willingness to give continuity to the industrial centre by favouring activities of service for businesses, such as those concerning logistics and distribution, to be established on the lagoon edge. Here, the context is characterised by some leading sectors and large enterprises; production excellence in some processes but avoiding extreme specialisation; and a fabric of small and medium enterprises. This is a scenario which could not be considered anything but desirable. However, in order for it to be feasible, immediate decisions and investments are needed.

4. The Port of Venice

A fairly important part of the Venetian production system depends on the role played by Venice's port and the opportunities this gives for the development of traffic. This very intermodality of transport is already one of the factors characterising the industrial centre of Porto Marghera and, potentially, will be an even more important one in the future. Furthermore, it is specifically sustained by both the possibility of transport on water and the sea access system.

The port activities as a whole can be divided into the commercial, industrial and oil sectors, to which passenger transport and pleasure craft tourism must be added. Over the last few years, the relative importance of these various functions has changed considerably. Comparing the tonnes/goods as the unit of measure for the first three functions, it can be seen that the resulting essential stability in total transport has been helped by a notable increase in commercial transort (nearly 75% from 1985 to 1995) which is balanced by the reduction in industrial and oil transport. A subdivision in terms of percentage, carried out in 1985, assigned 15.48% to the commercial sector, 36.85% to the industrial sector and 47.67% to oil. In 1995, these values had become 28.31%, 29.87% and 41.82% respectively.

The decline in industrial transport must be put down to the progressive closure of various manufacturing plants in the Porto Marghera area, while the continual increase in commercial traffic is attributable, on the one hand, to the gain of new transport (above all iron materials) and, on the other, to the consistent revitalisation of container traffic, the prospects for which appear much improved after the opening of the Gioia Tauro centre in the south of Italy.

With regard to the spatial dislocation of the Venice Port activities (with the exclusion of the areas of water, 90% of that which forms the port area has industrial and oil functions), it should be noted that a process of transferring all commercial activities to Porto Marghera has almost been completed, thereby progressively abandoning the areas of the Marittima in the historical centre (57% of the commercial traffic was concentrated here in 1950, which fell to 10% in 1995). According to the indications of the Port Authority's *Piano Operativo Triennale* (three year operational plan), these areas will be dedicated entirely to passenger traffic.

In terms of tonnes handled, bulk iron and steel products constitute the most important part both in terms of absolute value and of growth in the three-year period 1993–1995, in the same way as an important increase marked the aforementioned container traffic, by +25% between 1994 and 1995. A general increase in traffic was also witnessed in passenger transport in the 1985–95 period, although there were internal trends determined by the consequences of the Gulf War which influenced the whole tourism macro-sector. In 1995, the number of cruise ships which docked in Venice (to be considered as a home port, i.e., a port of passenger embarkation and disembarkation) was 201 with a total of 242,884 passengers. Ferry activity underwent a strong revitalisation in 1994 with the recovery of traffic to Greece, bringing the total number of ships to 338 and the number of passengers to 182,396 in 1995. Daily excursion traffic (hydrofoils and similar) which, as a result of its characteristics, has completely different impacts both on the port activities and on the income generated, must be added to this.

With regard to landing places, the Port Authority has designated the Marittima as the destination of cruise ships (together with goods traffic until its definitive transfer to the mainland), fast ships and daily ferries for Istria and Dalmatia are assigned to St. Basilio and St. Marta, whilst the most prestigious cruise ships are assigned to Riva Sette Martiri. Furthermore, in response to the dictates of the law for port reorganisation, the Venice Port Authority has decided to create a company which is expressly responsible for managing passenger traffic activities (cruise ships, ferries, others) following Genoa's example.

Considering that this is the general situation, its development increasingly depends on the necessity, which cannot be postponed any longer, to proceed with work on channel excavation. The delay in this has caused the rise in the channel beds which limits shipping traffic according to their draught. This essentially regards the tract linked with the sea which goes from Porto Marghera to St. Leonardo since the successive tract leading to the Malamocco inlet keeps its beds deep as a consequence of tidal movement. Today, it seems that the first steps are being taken but the delay has caused damage to the port itself as well as to all of its related businesses or clients. A study by the *Ente Zona* has estimated, for 1994, the damage caused by the lack of industrial channel bed restoration at 20 billion lira (lightened ships, ships with incomplete loads, hiring of ships of lower tonnage, ships waiting for the tide) to which another 10 billion lira can be added by estimating the damage to commercial traffic.

All of these problems have forced the port company to increase the costs of their services, thereby risking their own future since, at any moment, some companies could decide to move their transport to other ports or onto other types of transport. The Port Authority wants to act by investing in the possibility of 'bringing the production sector nearer to the port areas and reducing the transit times within the port structure as well as connecting the port area to the terrestrial transport network in a more fluid and rational way' (source: Port Authority). To do this, they set out some necessary action which, in addition to the crucial problem of channel excavation, includes:

- restructuring and strengthening the structures serving the commercial traffic in the Porto Marghera area;
- work regarding the practicability of and improvement of railway connections;
- revision of the structures serving passengers;
- valorisation of human resources and improvement of the working environment.

This is the scenario outlined by the Port Authority itself and is one which can only be supported; it works towards creating a structure which serves the production plants and which is able to receive and convey the various types of traffic (passenger included) which are vital to the entire economic system.

To conclude, the issue which is detrimental to port traffic is the depth of the Porto Marghera access channels. If their depth is excessive, there are significant consequences for the hydrological equilibrium of the lagoon and for its biological equilibrium and, therefore, for the protection of the entire environment. If the depth does not reach a certain threshold, both commercial and industrial traffic would be heavily penalised and would, therefore, have to move rapidly to other ports, thus causing the end of Porto Marghera as an area connected to the sea. A series of assessments need to be carried out with extreme honesty and transparency on these two apparently irreconcilable alternatives. Social processes, of which Agenda 21 is one form, should then be given the possibility of making a decision.

A brief analysis of container traffic is outlined in the next section. This tries to relate the probable evolutionary scenarios for this type of transport to the problem of developing the Venetian port and the problem of the depth of the beds.

5. Container traffic and the Venice port's potential

Since the 1950s, global container traffic has been characterised by a constant growth which has led to both growing attention from commercial operators and to the reorganisation of processes. Currently, some trends at the global level are fundamental for interpreting the changes taking place:

- the sector is highly concentrated: about 10 large ship owners control almost 90% of world traffic;
- the average size of the ships has increased considerably: today, ships with a tonnage of 6000 TEU are numerous whereas only 10 years ago the average did not exceed 1500 TEU;

- import-export seems oriented mainly to the final products of production processes (advanced intermediate products, compounds and finished products) rather than to primary materials.

Container traffic, specialised according to category, is more suited than others to transporting the former.

The United States, the Far East and Europe are the three markets around which the geography of container traffic is divided, and commercial exchange between them is always consistent. Due to their structural characteristics and geographical positions, the ports of Northern Europe are, clearly, the favoured destinations with respect to those of the Mediterranean. The latter were excluded from the main traffic until a few years ago, above all because of inefficiencies related to management (high costs, incomplete services, inadequate structures). Today, the difference between the Northern European basin and the Mediterranean basin has reduced notably as a result of, among other things, the individual action of the port companies.

The current routes are organised through direct or trans-shipping connections. When the first type is opted for, the voyage itinerary foresees stopping in one or more principal ports where the loading and unloading of goods is carried out (therefore, only the most important areas or those more involved in this type of traffic are used). In the trans-shipping connections, ships of large tonnage call at one or more ports of transhipment which are strategically positioned (and nearly always convenient from the cost perspective), and from which connecting feeder ships leave. These are of a smaller size and are able to reach other destinations (which cannot be otherwise reached with convenience due to their position with respect to the main route and to their amount of traffic due).

The average size of the new ships leads to very high management costs which can only be curbed by reducing the intermediate stops during the voyage to a minimum. The solution lies in the possibility of using the transhipment ports and, from there, the feeder connections. The creation of Gioia Tauro as a transhipment port creates new business opportunities for Italian ports. This is true for the ports of the Tyrrhenian Sea which can, nevertheless, also be involved in connections of the direct type. However, it is particularly crucial for the ports of the Adriatic, a corridor which almost always renders direct connections unfeasible and, instead, has greater opportunities for action in feeder traffic.

Venice is only involved in direct traffic through a connection with the Far East managed by ZIM. The rest is feeder traffic or light direct traffic, which is limited in both its dimensions and route. The ships used by ZIM for the direct connections reach a tonnage of 2500 TEU while feeder ships are rarely greater than 1500 TEU. The channel beds have made it possible to manage the traffic as we see it today, but the present depth (about nine and a half metres) is considered the minimum level required for not being excluded from the market. The costs for docking operations (tugboats, pilotage, navigation time from the roadstead to the docks, waiting time when there are tides) are also influenced by the characteristics of the beds. Therefore, Venice cannot effectively compete for some types of traffic with other ports such as Trieste and not even with Genoa in terms of areas to be served.

With respect to the current characteristics of container traffic, an increase in sea depth of about 60 cm would open up considerable commercial opportunities, by increasing total traffic by an estimated 30% (a part of which is to be taken from the Padua Interport). On the contrary, a reduction in depth of even only 50 cm would risk excluding the port of Venice from the market.

In conclusion, it must be stressed that the considerations presented remain valid only in light of the sector's current characteristics. The increase in the size of large ships could, in fact, lead to an increase in the size of feeder ships. In fact, by replacing the main ship with a bigger one, the ship owners favour converting world travelling ships to feeder ships rather than decommissioning them. If the average size of these feeder ships were to increase significantly, the whole debate on the suitability of the channel beds and, therefore, on Venice port's capacity to operate or not, would have to be revised.

6. The historical city, the islands and the mainland

The importance of the industrial centre of Porto Marghera in terms of employees, as well as its history and specificity, justify its previous individual treatment. It is possible to make other considerations for the area which can be called the residential area which comprises the remaining municipal territory.

Given its current design, the historical centre does not lend itself to the presence of modern industrial activities. As was noted in the first part of this chapter, this has not always been the case since, until last century, the old city and the islands which immediately overlook it were the sites of numerous and diversified industrial activities.[5] Island Venice is characterised, rather, by a high incidence of artisan activities which represent more than 50% of total material production in terms of local units. However, this figure must be correctly interpreted by considering that artisan supply is largely directed towards non-residents and must, therefore, be considered as a revenue of the commercial-tourism sector which forms the old city's production structure. In this way, instead of becoming a source of differentiation, artisan production reinforces the tourism monoculture.

Obviously, specific location factors influence the choice of other kinds of production plants. The site and supplying costs are very high and there does not appear to be a rosy future for those who direct production towards residence. For their part, the operators tend to exclusively assign the possibility of a revitalisation of the complete and differentiated production fabric to political choices in favour of the residents and small enterprises.

After all, the historical centre is characterised by the presence of few important entrepreneurial activities which, however, must be read as isolated endeavours rather than the tip of an iceberg. In this regard, the Arsenal shipbuilding yard in the historical centre, De Poli in Pellestrina and the growing Nautica Giudecca Centre can be mentioned. Some firms can be added to this, above all, those concerning construction. These are of a significant size and have their headquarters in the Venetian historical centre although the range of their activities extends well beyond it. According to the Industrial Association, the same opportunity for focusing on the advanced tertiary sector, thereby concentrating on immaterial production, does not seem capable of taking off. Currently, it is difficult to envisage scenarios capable of overturning or, at least re-balancing, the historical centre's production scenario. Among the other material production specialisations in the residential area, glass manufacturing on Murano, the shipyards and a vast array of minor artisan activities in general all stand out, the most well-known of the latter perhaps being lace-making on Burano. These are very particular cases of historical production specialisations which contribute significantly in characterising the city's economy.

In the historical city and islands, artisan activities prevail with medium-sized Local Units which are rather limited in terms of employees, with the exception of a few important and known production structures, such as the production district of Murano. This has 2303 employees in material production: the ratio between employees involved in such production: resident population is 0.42 and it is, therefore, an important production district in absolute terms. In particular, in terms of the manufacture of blown and handmade glass, the famous typical local production which employs almost 65% of the employees of material production on the island, the operators maintain that an immediate effort is needed to improve the relationship with the outlet market, in terms of both distribution and communication. However, among the dangers which most threaten the island's production activities, it is the change of generations which risks threatening the very survival of the sector.

Nevertheless, in general terms, the question of the development of material production in the residential areas of the Venice municipality, particularly in the lagoon parts, must be linked to several large issues which have been discussed for a long time but for which it seems difficult to find solutions quickly. These are the protection of the territory, transport and access systems, residence. Many of the production activities related to residence have ceased, having followed the trends of the exodus to the mainland. It is unlikely that new enterprises would be established in the lagoon, since they have to take on the inconvenience and costs related to transport which, in fact, seriously affect performance (interruption due to sea/land transport). The situation on the mainland appears different. Here, on the contrary, a series of activities are already underway. However, it is the entire municipal area which needs to be looked at if the imbalance, which has often been considered undesirable, is to be avoided.

7. The principal sectors of material production: essential considerations

The immediately obvious aspects of the production structure of material production[6] (MP) in the Venetian territory can be summarised as follows:

- The 1991 census data reveals that, in terms of contribution to employment, material production in the Venice municipality equals 32.3% of the total.
- The leading role in this is played by the medium-large industrial activities; the small enterprises, although numerous, are comparatively less significant in terms of the capacity to generate employment.
- The different areas of the municipal territory show marked diversity in terms of the structures' size.
- The distribution of MP in terms of destination of use shows that three-quarters is attributed to goods destined for production and one-quarter is destined for consumption.
- Over 57% of the production activities are of the land-consuming type.
- In spite of the fact that radical changes have been seen in many of the material production sectors and further ones are anticipated, which often lead to the re-dimensioning or the abandoning of entire portions of the production process, the system as a whole appears to be characterised by a strong resistance to change.
- The area studied is characterised by a series of production specialisations.

In an attempt to identify the leading Venetian MP sectors, in terms of their degree of production specialisation and their contribution to the workforce employed in this type of activity, the following sectors can be identified:

- ship-building;
- oil industry (refineries and deposits);
- petrochemical industry;
- aluminium intermediate products;
- electricity production;
- artistic glass production on Murano;
- construction and building industry.

Ship-building is, perhaps, the sector which combines tradition and specialisation with production excellence and market success more than the others. The focal point of the Venetian ship-building industry is constituted by Fincantieri, a company which operates in Venice with about 1300 employees and has asserted itself as a world leader in the production of cruise ships.[7] The possibilities for developing this company's Venetian activities further are linked to the urgent need to excavate the channels in order to deepen them from their present six metres to the necessary eight metres. The problem is evident, above all, in light of the trends in constructing passenger ships of increasingly greater size. At the moment, the company is forced to transfer the hull to Monfalcone in order to fit out the ship and to launch it, with the obvious loss of both related revenue and of opportunities for local companies involved in the construction and the fitting out of ships.

Nevertheless, the Venetian ship-building industry as a whole, despite fluctuations (the crises of the Toffolo and Lucchese shipyards as well as the Arsenal shipyard are recent) is still a promising sector which has other ship-building concerns which deserve a mention. Among these, the De Poli shipyard in Pellestrina, which is strong due to its advanced technology in processes and products, is specialised in multi-use tankers capable of transporting numerous types of loads, even corrosive and dangerous ones. The Venice Arsenal is the headquarters of the company with the same name which operates using a large area of the whole and the historical dry dock basins for ship repairs and reconstruction. Before the recent crisis, this was also supposed to host the new mixed operational unit (public-private) which was primarily for work on the ACTV fleet and other potential commissions. Finally, the islands host smaller shipyards aimed at the pleasure craft and nautical services markets. The Nautica Giudecca Centre, an entrepreneurial initiative which has been set in motion with co-ordinated investments (private and public), stands out among the emerging concerns. The beneficial effects of this on the local economic structure will range from the recovery of urban areas destined for public use and of buildings of architectural interest, to the creation of new enterprises, to giving new life to what already exists in a sector which has its roots in the Venetian socio-economic fabric.

The oil sector is associated with the Agip refinery (about 3.5 tonnes/year) and with the seven coastal deposits (about 10 million shipments a year) belonging to seven different companies. Excluding export, about 5 million tonnes a year remain in Venice compared to an average demand in the Northeast estimated at 6.5 million tonnes. As the refinery at Ravenna is closed, the nearest plants are those at Mantua and Cremona. Therefore, the importance of the oil centre in Marghera for the whole region is evident. Today, the sector has about 750–800 employees, the majority of whom work in the refinery.

It is unnecessary to stress that this is one of the sectors at the centre of discussions on the environmental problems blamed on the existence of the Malamocco-Marghera channel (the so-called oil-tanker channel) which are derived from tanker traffic. All of this has led to the presentation of a modification to the regulatory plan for Marghera which originally defined the refinery area as 'oil activity in an inappropriate area', and which, after some pressure from the area bodies and Assindustria above all, changed the phrase to '*D/V* area (destined to be a green area) – oil activity in operation, destined to be an urban green area for environmental restoration', and (with the introduction of article 28 bis), in fact, foresees the maintenance of the refinery's activities and those of the other existing oil plants with the possibility of implementing all of the necessary intervention for adjustment and conversion (transformation into a green area will only be carried out in the case of total or partial decommissioning of the oil activities).

The petrochemical industry is also a special case which depends on Enichem and other related industries which are mainly employed in PVC production. Serious problems related to the dangers of pollution have led to significant intervention in the plants and in the production system. In this case, in addition to the issues related to the plants' safety and respect of standards, there is also the estimation of the economic advantage which has confirmed that it is no longer profitable to manufacture certain goods in Italy (therefore, the problem is not confined to Porto Marghera) since it is possible to transfer the activity to third world countries with lower energy and labour costs. This has led to the closure of some companies (Enichem agriculture for fertilisers) which, together with some transfers, has redesigned the order of the sector in an obvious way.

With regard to the aluminium sector, while the non-profitability of primary material production is clear as it is characterised by high electricity consumption and has, therefore, been penalised since the period of the 1970s oil crisis, instead, intermediate aluminium production remains efficient. In Porto Marghera today, this production is represented by the plants and activities of Alcoa, which has taken over Alumix for lamination and extrusion, and by the drawing processes which are still Alumix's concern. Alcoa's arrival is a concrete example of a large international group's interest in the Venetian industrial centre and a model on which it seems opportune to focus. To do this, however, conditions of stability and transparency, which allow decisions founded on estimations of economic benefits for the companies which want to invest, need to be guaranteed.

The electricity sector can also be considered a leader. The four plants in the sector have about 750 employees and have a total output which is sufficient for satisfying the demand of the local and surrounding areas. Of the companies to be highlighted, Edison can be placed alongside Enel in terms of production. Sales remain the exclusive competence of Enel but other companies can produce electricity to then pass it to Enel or satisfy their own needs with it. For example, the Agip refinery uses fuel oil for their own production of electricity, although for reasons of security it maintains a relationship of integrated supply with Enel.

The production of artistic glass represents an element which unites a historical-cultural root of international importance with an employment capacity and highly profitable production capacity. The characteristic element of this industry is a virtuous mixture of uniqueness, high value added and a strong relationship with the values and competence rooted in the local population. Among the suitable initiatives for the valorisation of this sector, there are no elements which prejudice its vitality, with the exception of the generic danger of damaging the image of Muranese products which derives from the circulation of imitations and, in general, from a potential decline in the quality of the products presented in the guise of Murano glass.

Finally, the construction sector provides a contribution to employment which, as well as being significant in numerical terms (over 7000 employees found by the CCIIA census in the first half of 1996), also seems to evolve in contradiction to trends (23% more than found in the census of 1991). It is reasonable to presume that if this figure, on the one hand, naturally accompanies the existence of a densely populated area with a high rate of urbanisation, on the other, it is at least partly sustained by the frequency and importance of the ordinary and extraordinary maintenance works on the buildings of the historical centre, which are, without a doubt, greater than in an ordinary urban centre.

The above listed sectors all provide a significant contribution to employment. In this respect, it should be noted that, from 1991 to the present, there are no sources which supply data comparable to that gathered by the census which makes it necessary to use data from sources other than Istat. For this reason, the comparison made here uses data that is not perfectly homogeneous. Nevertheless, the information can be considered significant, at least in terms of trends. In this sense, it is worth stressing that what emerges from recent data does not substantially change the picture of the MP production specialisation as outlined with the census data.

The total number of MP employees recorded by the Chamber of Commerce archives registers a reduction of about 20% with respect to the 1991 figure, with a loss that can be estimated at just under 7000 jobs. This reduction in personnel concerns nearly all of the MP sectors, with a few exceptions. The construction sector has grown by more than 1300 employees, that of fuel sales and commercial vehicle repair would appear to have about 400 employees more in 1995 than in 1991, while employment in the sectors of metal production and wholesale trade has increased by about 330 units. In addition, there is a percentage increase in the number of employees in the clothing industry, wood industry, paper industry and in employees in car production and other manufacturing industries, but the increase in terms of units is modest.

In contrast, all of the other sectors decline, starting with the chemical industry which has lost nearly 1900 employees. Moreover, this is a trend which has involved the entire sector throughout the country and has been mainly generated by a significant fall in internal demand and the coincidence of this phenomenon with the processes of capitalisation and externalisation of the production of basic chemicals in third world countries (Guelpa, 1993).

Metal manufacture has also declined in terms of employment (over 1100 employees), along with mechanical sector and the food industry (both by more than 850 jobs), while the oil and coke industry has lost a total of 700 employees. Even with all the caution which the case deserves as a result of the aforementioned diversity in sources, the tendency towards reducing the use of the workforce within the process of industrial transformation is a general phenomenon in all industrialised countries. For the sake of comparison, data related to the employment trends in some of the industrial macro-sectors in the areas neighbouring Venice, more precisely in the Provinces of Venice, Treviso and Padua, can be looked at. In these provinces, save for few exceptions, the phase of employment growth in the principal industrial activities finished around the end of the 1980s, while from 1991 to the present employment has generally diminished. The singular stability maintained in the mechanical sector in Venice and Treviso and in Treviso's heavy industry is the exceptions to this trend.

Employment in Porto Marghera has paid heavily for this overall contraction. Data referring exclusively to this area, show that between 1991 and the end of 1994 the number of workers fell by 5358 units. The sectors in which the most notable losses are seen are those which have already been mentioned with reference to the municipal data.

In spite of this negative trend, the Venice municipality still maintains a significant production characterisation in these very sectors which have recorded the greatest losses. The data which permits us to sustain this comes from the comparison between municipal energy consumption, subdivided according to each activity, and that for the Veneto region, a comparison from which the preponderant importance of the aircraft activities (99.8% of regional electricity consumption), of the oil sector (96.4% of refining and storage), of coke products (83.5%) and aluminium (96%) powerfully emerges.

As a whole, if, on the one hand, the total picture makes all of the problems related to production conversion, which was broadly discussed in the first part of this chapter, stand out, it is also true that the data available until now would suggest that a process of restructuring and conversion is taking place rather than one of decommissioning. In fact, the data related to employment clearly reveals the technological perfection of industrial processes which tends to increase the system's productivity with, however, the disadvantage of lower employment. This is an aspect which suggests that the assessment of the problem related to the area's production conversion would need to be accompanied by medium- and long-term projects for the qualification of the workforce. This is a very complex task but the Venetian University can have a prominent role in accomplishing it.

The theory which maintains the importance of material production's role in the city is one which has been expressed many times. The main problem remains that of managing to outline a scenario which is characterised by a relative equilibrium between sectors, impact on employment, areas for locating production activities, revitalisation of residence, respect for the environment, quality of life.

Among the principal steps in this direction, it is held that the search for production diversification must be included, thereby avoiding the risk which any marked specialisation poses to an urban system. Evidently, all this must take place without penalising the sectors which, more than others, have represented and can represent in the future the nature of the Venetian production system.

The complex subject of residence appears among the hottest issues for the future of the island part of the city and, therefore, for the area's economy. Among the sectors traditionally linked to a residential-type demand, bread production is exemplary. The opinion on the historical centre expressed by those responsible for the industrial association and the Venetian bakery consortium, sketches a picture which is not very reassuring. In terms of production units, no changes have been recorded, but the number of employees has diminished considerably over the last few years. Average daily production proves to have halved as a consequence of two kinds of factors, the national trend of a general reduction in the average daily consumption and, in the specific case of Venice, the exodus to the mainland.

During the 1950s and 1960s, 95% of production was destined for the resident population and 5% for tourists; today the trend has been reversed, with only 20% destined for the residents. This reversal (given that the fall in the absolute value of production remains the same) is considered particularly risky, considering the uncertainty and seasonal nature of tourist flows which make the management of the activity problematic. Moreover, transport costs have a decisive effect (40% of the cost of the primary material), making it even more urgent to find a policy which, by encouraging residence and the activities of small production units serving residents, revitalises the sector.

8. Suitable policies

As mentioned at the beginning, the so-called process of de-industrialisation which has struck the more developed countries does not, in reality, imply a reduction in the value of MP but, rather, a reduction in employees. This occurs because of the great increase in productivity in all sectors. Therefore, the hypothesis must be considered that MP in the Venetian area has to be reduced very rapidly until it ceases.

Over the last few years, it has often been said that the Porto Marghera area should 'turn over a new leaf' almost completely and become an area of new production plants which are no longer polluting, nor basic in nature, and should provide enough space to assign to non-material activities (research, planning, production of prototypes, the preparation of new enterprises, experimental and wider application of information technology, brokerage etc.).

While there are no negative side-effects as long as part of the areas available are used for such activities (the first part of the technology-science centre is already in operation) it is, however, reasonable to believe that these activities could also be carried out elsewhere. Therefore, the objective of having a radical transformation of this industrial area over the next ten-twenty years, with the abandonment of the material production that characterises it, does not appear very realistic today. Rather, the possibility of the decisive implementation of two types of intervention could be considered, both of which, however, require significant financial resources: (a) investments in technological innovation which could make the production plants sited there more competitive and convenient since they can enjoy specific location advantages; (b) investments in environmental safety (the micro- and macro-environment) in order to make it practically impossible for serious accidents, which affect the lagoon and the Venetian area in general and, particularly, in the workplace, to occur.

The technology available today allows these objectives to be reached. Moreover, these elements would have to be included in the debate taking place on Porto Marghera before assuming *a priori* a position against the maintenance of certain types of industry.

If the way to maintain such types of industry is through a series of targeted investments, it remains to be seen whether the sum of the investments made by each firm operating there can be paid back at an acceptable rate of return, but this is a problem of calculating the advantage and making strategic decisions on the part of the firms. Such decisions which, as said, entail enormous investments and, hence, the undertaking of some risk, can only reasonably be taken if the uncertainty and, therefore, the complexity is reduced to some extent. This could be carried out if the Public Administration, at all levels of authority, agree on a clear course of action, determine the boundaries and focus on credible scenarios which can be depended on. Otherwise, in the uncertainty of the context, firms will rightly postpone investments and will gradually search for solutions which are less problematic and uncertain with respect to their strategic plans. What counts for the decisions-makers, whether private or public, is to deal with scenarios which are neither excessively vague nor uncertain. A clear 'no' to certain types of production is much better than a continual and confused chatter in which ideological positions, cultural delays, and the search for compromise at any cost are fused.

This last consideration obviously applies even more for firms which would potentially like to establish new production activities in the area since, in this case, the barriers to exit operating for those activities which have been installed for some time would not be in effect (sunk costs).

In any case, alongside the material production activities which have survived the slimming down of Porto Marghera, other production activities can and will be able to locate there, but it is appropriate to distinguish two models when referring to these. There are the firms which operate with activities which consider the direct connection with the sea (the availability of equipped docks) a strong and decisive motive for their potential choice of location (the first type). There are also the firms which, due to the nature of their activity, could locate elsewhere, but are unable to find sites which are acceptable to the local population in our region since they are not sufficiently separated from the residential area and are in areas which are not equipped with the specific infrastructure of an industrial area. 'With a Northeast which is territorially saturated with industrial plants, Marghera can still present itself as the winning card of Venetian development, due to the strength guaranteed by the presence of the port, airport, railway and road infrastructure and as a result of the advantages derived from the large number of other existing professions and the entrepreneurial mentality generated by long-standing industrialisation'.[8]

Furthermore, we must remember that Porto Marghera also appears important because of its commercial port functions, the majority of which, as noted, are already being carried out in this area but will be completely carried out here in a few months, as soon as the transfer of the goods traffic from the Marittima (in the historical centre) has been fully implemented.

There is a possibility that Porto Marghera could remain a centre of material production with an importance which extends well beyond the Venetian area, which could, therefore, contribute to its export base without damaging the area from the environmental perspective, but, instead, by ensuring a production equilibrium between material and immaterial which seems to be the best solution from the structural point of view. This possibility is linked to a series of conditions which can be summarised as follows:

(1) Both the firms which operate there and the other public and private actors linked to it have to make the investments which are commonly recognised as necessary for reducing the likelihood of damage to the environment, in its widest sense (natural, social, human), to negligible values.
(2) Favouring areas which have become available over the last ten years which would only be suitable for industrial use with the implementation of a plan for restoration.
(3) The clear definition of context conditions (the depth of the access channels, maximum emission values allowed, etc.) and of the urban development plan not only for Porto Marghera but for the entire Venetian area in order to supply suitably founded elements on which to make investment decisions which are often of remarkable sums (to the order of hundreds of millions of lira for each firm).
(4) The identification, by entrepreneurs and managers, of a certain number of entrepreneurial formulae (organisation, product, market, technology), on whose success in the medium- to long-term in Porto Marghera it would be worth investing in and gambling on.
(5) The growth of a production culture in Venetian civil society which admits and nourishes one of the riches of the area, that is, the existence of very different production and cultural activities which, if revived, are able to provide a valid defence against the supreme development of the monoculture of tourism. The latter allows for rapid and relatively easy earnings with limited investments and would become excessively risky for Venetians if they, directly or indirectly, concentrated the majority of their production activities in it.

At this point, however, some questions must inevitably be posed:

- considering the vastness of the intervention and the financial resources necessary, is it really advantageous to bring the industrial area of Porto Marghera up to full environmental compatibility?
- what could be the value of industries which could locate *ex-novo* and of those which would be able to continue operating with respect to the costs of restoration of the terrain and to the investments in plants and infrastructure?
- in what other conditions is it economically possible to maintain Porto Marghera as an area of material production?

It is not easy to answer these questions without a sector by sector analysis of the potential and the relative location advantages, a task which is impossible to carry out here.

In short, it is our conviction that two problems must be resolved in order to determine the future of the entire area. These are, more precisely: (1) the compatibility of the material production activities present there (oil refining and basic chemicals, above all) with the rest of the Venetian area; (2) the compatibility of the port operations with the lagoon environment.

Together with other issues such as, for example, the quantity and quality of tourist-based functions, these embody the alternatives which can be practically formulated and, therefore, the scenarios which can reasonably be envisaged for the future of the Venetian area and the Venetian historical centre in the medium- to long-term.

9. The directions of sustainable development

It must be said that if Venice has witnessed a succession of unrealised projects for a long time, this is not only due to the fact that no room has been given to the technical effort necessary for their achievement, but also because, by paying attention to large interventions and to the most macroscopic impacts, the invisible economic and social micro-dynamics are often ignored. These are often slow but incredibly resistant and disruptive in the long-term and tend to oppose certain types of processes.[9] Among these, two appear to be of great importance: trust and socio-cultural acceptability.

The mistrust in the credibility of a radically innovative and beneficial change risks threatening the participative process of Agenda 21 at its root. The lack of realism (when not of good sense) in the projects which aim at revolutionising the urban and socio-economic[10] fabric of the city, needs to be dealt with by incorporating credible, slow and socially acceptable tendencies; recent experience clearly shows the path to follow in this respect. For example, after many years and only recently, in Venice and Mestre work aims at improving the urban landscape, redesigning the road conditions, excavating canals in the historical centre and at maintaining buildings and coastal defence. These, and many other projects, which, individually, appear modest in comparison with some great plans which have fallen through in the past, reinstate the feeling that the inhabitants' quality of life can improve in an ordinary way. It is this very road of realism which, in our opinion, must be followed in order not to risk losing the trust which sustains the participative process.

Socio-cultural acceptability of a project is meant as something different to social consensus. The latter is prevalently the expression of interest groups and, therefore, assumes importance based on the power which each group is able to express through institutional channels. Instead, socio-cultural acceptability is formed by the subjective perception of the nearness to ones own values. It is not translated into presentations or in Council meetings, but rather into shifts in opinion or into widespread ill-humour, as well as into demonstrations. A project which does not enjoy this acceptability, whether it is valid or not, is destined to be considered a failure (e.g. Expo) or to be perceived as an act of collusion (e.g., the extraction of methane in the Upper Adriatic). Such a subject could appear banal if it were not for the fact that the recent past is full of projects and results which give no sign of attention being paid to these two conditions (e.g. Dekleva, 1972).

In short, a possible key for reading this chapter is that of opposing the myth of the technical measurability of development's sustainability with the capacity for carrying out an analysis which is able to provide information on the actual state and the possible evolutionary trends for the actors who are responsible for solving the real problem. For this reason, the first step in constructing a sustainable development scenario consists in the effective translation of the concept of sustainability, thereby identifying the prerequisites that a evolutionary course must have in order to be considered as such.

10. Making the concept of sustainable development operational

The concept of sustainable development involves two fundamental dimensions, each of which is strictly related to the other. The first dimension is the durability, which is meant as the process's capacity to nourish and replicate itself. The second dimension is the desirability, which is meant as the existence of a wide social consensus on the effects of the process. This consensus is achieved when the process has the capacity to satisfy, by best reconciling them, all of the possible needs presented in a present-day civil society, i.e. the environmental, economic and social needs. However, a complication emerges. If on the one hand it would be theoretically possible, in terms of isolated aspects, to single out the desirable directions of development, at the stage when the interrelationships between the various aspects have to be considered, it becomes very difficult to express an unequivocal judgement on the most opportune direction. For example, it is difficult for an indicator to give an indisputable answer on the sustainability of a trend with regard to the different aspects which are involved. An indicator of productivity, for example, can be interpreted differently according to whether it is considered in economic terms or in terms of employment. For this reason it is, above all, necessary to reflect on the relationships between economic sustainability, socio-cultural sustainability and environmental sustainability. More precisely, since universally shared criteria for the desirable form of socio-economic development do not exist, it is necessary to formulate hypotheses on this subject. Care must be taken to ensure that no scenario exists which is not founded on hypotheses about the dimensions of sustainability. In any social science it is utopian to think it possible to create models which can provide an evolutionary framework which leaves the starting hypotheses of the model itself out of consideration. Furthermore, to omit the explanation is an unacceptable methodological choice.

Equally, starting up a debate on the configuration of the scenario would be in vain since it risks falling on common ground where it is not possible to disagree. Indeed, the ideal scenario is one which durably and simultaneously achieves the elevated production of wealth, large quantities of individual free time, comprehensive distribution of wealth, absence of pollution, and so on.

There is nothing else to say about such a scenario except that it is optimal and currently desirable to the same extent as it is unachievable and, therefore, useless even if only seen from the perspective of an ideal-typical reference.

This is why, if on the one hand the debate must instead centre on hypotheses of sustainability and on decisions on what is held to be most desired (looking not only at that which is obtained but also at what, on the contrary, is renounced), on the other hand, the contribution of the analysis, which must be made in the technical phase, is to identify the variables which act on the socio-economic system and to identify their interaction, to define the instruments for guiding the direction of development and to possibly collaborate in drawing up and executing the plan.

To conclude, making the concept of the sustainability of development operational for the Venetian territory means no more than expressing the hypothesis at the basis of the scenario which is chosen for proposal on the basis of it being the most reasonable among all those which can comply with the needs of the various social components of the city. In particular, specific attention is given to material production here. Nevertheless, for the reasons given above, it is not possible to talk about sustainable development by accentuating the differences between the different plans for socio-economic development. Therefore, the comments made here deal with a complete overview of the problems of development.

10.1. Hypotheses of sustainability

In an area such as the Municipality of Venice which connects a peripheral part, which is highly interconnected with the Veneto hinterland, to a historical centre and to a lagoon component which both remain relatively isolated from the rest of the territory, the existence of an economic system which is differentiated in terms of sectors, production forms and the size of production unit, is both the symptom of and the driving force behind the wealth and welfare (in general, the quality of life) in socio-economic and cultural terms.

Even if production specialisation has advantages in some respects, these advantages risk prejudicing the quality of life of the population of the territory when the displacement of other production activities, and their related socio-cultural heritage, occurs. A phenomenon of this type is visible in Venice in the case of tourism, where its current form of development has triggered a series of phenomena which inflict very great hardship on the residents in numerous ways; high prices, problems of urban mobility, unlawful activities, petty crime, etc.

Along with the socio-cultural phenomena, it is also necessary to consider that an excessive production specialisation in a restricted area can also have negative effects on the economy itself, since potential effects of recession (both from cyclical downturns and trend recessions) can be tolerated better by a differentiated economic system. This is because, when other conditions remain equal, it is less likely that the crisis affects non-interdependent sectors of activity in an equal manner. As a result, the non-penalised sectors can act as an element which triggers economic growth (in the case of phenomena that tend not to be repeated) or can play a counterbalancing role (in the case of phenomena which are seasonal and related to economic trends). Finally, in the case of a recession, a potential policy of conversion and/or revitalisation would prove to be much less burdensome on the social and economic level and the influence of activities involved in the process of restructuring/converting would be less pronounced on the local economy.

This kind of argument could be countered by the fact that specialisation could guarantee a competitiveness which is useful for overcoming crisis situations. Nevertheless, this interpretation is based on the supposition that specialisation is synonymous with having a distinct advantage. The chemical industry in Porto Marghera is an example of how things can develop differently with the passage of time.

If this is true, a hypothetical plan for the programming and development of production activities should be able to set out the advantages of specialisation, as well as those of diversification, in order to take advantage of the opportunities inherent to a local system. These can be summarised as:

(1) the existing competence and structures,
(2) the natural vocations related to the local culture,
(3) the territorial and location availability,
(4) a boost from inter-sectorial relations.

Furthermore, the idea of sustainability requires the respect of certain constraints:

(1) compatibility between the different possible types of activity and polymorphism of the production structure,
(2) adequate levels of employment and economic growth for the local population,
(3) respect of local culture and prospects for growth (competence, innovation and research),
(4) respect for the environment.

In the following, an attempt is made to tackle some of the many aspects related to this subject, in the full knowledge, however, that this is not the place to examine them in suitable depth.

10.2. Development Opportunities

Existing competence and structures

In terms of material production in the Venetian territory, specific competence and structures constituting a strength, which is difficult to reproduce in the short-term, are widespread. The sectors identified in this respect are: the manufacture of aluminium, the oil industry, the petrochemical industry, the shipbuilding industry, more minor shipbuilding, medium and heavy mechanical industry, the works on the lagoon and sea, the manufacture of blown and hand made glass. There is also a port in evolution, the economic impact of which could be exploited further with appropriate interchange structures for goods transport.

Natural vocations linked to local culture

The so-called traditional activities, as well as being closely related to the residents' quality of life (see the constraints), can be the sources of opportunities for at least two reasons: employment of marginal groups in the workforce and boosting demand for other goods and services.

The first aspect regards activities such as the manufacture of glass 'a canna', the lace making on Burano, woodwork for boats and decoration, furniture restoration, mask making and others. These are all activities which, as a whole, have employment and revenue advantages which are, perhaps, of little relevance in percentage terms but have the merit of employing (or being able to employ), on a full- or part-time basis, groups of the population (mainly the young and the old) who would otherwise remain inactive.

The second aspect, boosting the demand for other goods and services, refers to the fact that certain marginal activities, mentioned above, can have connections (in terms of image or function) with more important activities. The connection with tourism is obvious, but it is also possible to envisage a scenario in which, for example, the presence of diverse forms of typical crafts in the historical centre, located at the same place of sales (the typical Venetian store), can also encourage the inhabitants of the metropolitan and mainland area to visit the old city more frequently. As well as employment benefits and variety in the urban production fabric, this also creates a reason for the greater mobility of residents.

Territorial and location availability

The great diversity of environments, in the term's widest sense, which are rooted in the Municipality of Venice must be regarded as a potential source of opportunity rather than a point of structural weakness. This is particularly true where particular advantages of location specialisation can be exploited.

Porto Marghera is an area which could be put forward for employment and production revitalisation on the condition that the real costs and the programme for the area's use (in particular, the legislative reference framework) are made clear. Development linked to immaterial activities in the form of a centre for research and innovation can also be envisaged. However, it is also important to highlight one aspect: although it cannot be disputed a transformation into forms of production which respect the environment and are linked to innovative and developing sectors is desirable, it is also obvious that entrusting the transition to the inevitability of decline and to market forces, that is, to let the trends continue without adequate planning, firstly means squandering an important heritage of experience, knowledge and technology, and secondly, means the impoverishment of the economic structure of the

Venetian territory. It could not be justified, for example, if once the industrial areas of Porto Marghera were restored, or at least, were made to act within the regulatory requirements, at the expense of the collective, they then yielded to activities which did not have a positive effect on the economy of the collective. Equally, letting the current state of decline mature to the point where any alternative solution to its abandonment appears as an improvement is also unacceptable. At the moment, location advantages for the activities of industrial transformation in Porto Marghera stand out unequivocally; an advantage which mainly lies in the following factors: isolation from the residential sphere (already operating); limitation of environmental impact (already existing in part, in part to be achieved through appropriate investments at both the level of the individual firm as at the general area level); infrastructure such as the road and railway networks and dockside (already in operation and more to be executed); finally, a consistent supply of a potential workforce settled within a limited range in terms of travel time. Nevertheless, at this stage these advantages remain ineffective since the basic uncertainty about the costs and comprehensive programme for revitalisation tends to frustrate the evaluation of the economic opportunities for investment in the area.

The city waterfront is the site preferred for minor boatyard activities, while certain areas of island Venice and the shores of the lagoon represent the same for shipbuilding. The reputation of companies such as Fincantieri and De Poli is well-known, but other lesser-known cases exist which aim for or have achieved ambitious goals. In particular, the smaller boatyards on the islands are partially penalised by their location, but this could also be a source of opportunity if internal demand were fully exploited and synergy with the demands of tourism were sought. In this respect, the role of local demand in acting as a trigger to economic growth should not be underestimated. This demand can be fed by the mass of works required for maintaining the various parts which constitute the city of water, and by the demand for working or supply boats or their related maintenance and service activities. In particular, it should be considered that once the works which appear to be extraordinary, such as those in the lagoon environment, on buildings, on the city canals, etc., begin to operate, in reality they take on a permanent character, thereby opening up possibilities for construction activities and for the development of technology for marine intervention.

These are elements in the production district of blown and hand-made glass on Murano which contribute to the definition of a production district in relation to the Muranese artistic production already in a historical perspective. The forced transfer of the glass factories, decreed for reasons of safety by the Great Council at the end of the XIII century and, as a result, the rigid institutional control on the state of the art and on its form of transmission has engendered the formation of a selective socio-cultural fabric which is strongly directed towards the self-preservation of roles and competence. On the one hand, this has guaranteed the endurance into our times of manufacturing techniques which are unique in the world. On the other hand, the Muranese district's capacity for communication with a changeable reference environment such as today's appears limited and it appears more oriented towards involution rather than towards entering the competitive arena. The possibilities of valorising this sector further are certainly numerous and principally related to the perceived high differentiation of these types of products.

The lagoon and the islands are already used for the intensive exploitation of activities such as fishing and fish-farming. Unfortunately, the unauthorised practice of these activities threatens to irreversibly destroy a highly valuable resource which should be valorised and regulated instead. On the other hand, the Cavallino coastline, the island of St. Erasmo and the Venetian mainland host a significant amount of agricultural activities. In fact, there are over 19,000 ha in the Venice Municipality which are destined for agricultural use, with employees estimated at about 2400. An important fact should be noted: the development of agricultural activities in the cultivated lagoon islands (St. Erasmo, Vignole, Mazzorbo and Torcello) and on the Cavallino coast are developing in accordance with innovative models. However, they do not damage the landscape as a result and they are partly able to limit the environmental impact (Benevolo, 1996, pp. 36ff.).

Boosting inter-sectorial relations

The activities already existing in the Venice Municipality can be sources of further economic development once the manufactured goods produced by them find favourable conditions for developing inside the same Municipal territory. In particular, the availability of area for industrial use in Porto Marghera should be exploited to allow production activities, which have the same process as those already present, to locate in proximity to them. In order to identify the most suitable instruments for ensuring that such a boost to activities is facilitated, it is necessary, among other things, to have an exhaustive picture of the opportunities for locating in the rest of the Province and/or the Region, by gathering the information that a businessman would use to decide on the most opportune location for his company. In this respect, the location advantage of the Venice Municipality in terms of its ability to transport goods must not be underestimated, just as the overall increase in the transport of goods between 1994 and 1995 seems to indicate (+17.8% for the railway; +14.6% for the airport; +9.6% for the port, divided into: commercial +14.6%, industrial +11.1%, oil goods +4.4%, container +11.6%; +10.7% for motorway toll booths from and to Venice).

10.3. Constraints on Sustainability

Equilibrium between the different possible types of activities and polymorphism of the production structure

In brief, as has already been mentioned, the main advantage derived from the presence of a differentiated production system is the reduced impact of a potential sectorial crisis on total local employment. In addition to this aspect, there are the considerations inherent to the social and cultural desirability of multiple forms which oppose the dominance of one or few models. On the other hand, the most obvious negative aspect is the absence of specialisation advantages, which can be translated, among other things, into lower competitiveness in terms of cost, reduced capacity for innovation and reduced possibility for synergy in terms of image.

The intermediate course consists in giving priority to the points of strength without penalising or displacing initiatives of another nature in doing so, possibly operating in such a way as to avoid there being excessively marked differences in opportunity. The case of tourism is emblematic: relatively few barriers to entry, little regulation, few inspections, high profit margins and a quick rate of return are the foundations for the success of tourism as well as the consequent displacement of other business initiatives.

Adequate levels of employment and economic growth for the local population and services for them

It is necessary to ensure that the resident population can gain access to the benefits derived from the local economy's development. The profits generated by activities located in an area do not automatically go to the local population. In the same way, it cannot be guaranteed that the opportunities and services which render a given environment favourable for the creation of enterprise also bring advantages to those residing there. In a sense, it may even seem rhetorical to assert that the opportunities of a territory must be exploited, within their possible limits, for the benefit of the collective living there, but, looking closely, this does not always occur. For example, if the areas which have the greatest landscape value in the historical centre and Lido are looked at (the Biennial Gardens, the head of St. Elena island, the eastern area of the Giudecca, the Arsenal, the island promontories of St. Nicolò and Alberoni, the Lido coastline, the islands of Vignole and Certosa) it can be seen that the activities established there are aimed at a rather limited collective and they only partly bring benefits to the residents (Art Biennial, Naval College, Finance Police, Congress Centre, Military Marina, bathing concessionaires, Airport, golf club, military zone, etc.). Saying this is not meant in anyway as a judgement on the activities' current location. Instead, it is meant to provoke thought on the extent to which the Venetian territory can be enjoyed by the resident public, especially if the different capacity of these activities for contributing compared to part of the tourism sector is considered. In other words, it is necessary to be vigilant in order to avoid creating a systematic migration of development benefits (in its widest sense) which are beyond the citizens' reach. This phenomenon is also seen in the retail trade network which is decisively oriented towards the visitors' demands rather than those of residents.

Respect for the local culture and for the prospects for growth (competence, innovation and research)

It is important not to distort the values of the resident population. Firstly, this entails respect for shared principles and giving incentives to individual initiatives which can play an important role in determining the residents' quality of life. Indeed, even if the impact of these initiatives in terms of employment and profits are marginal, they are responsible for transmitting values and for creating a relationship of continuity between generations. In addition and more generally, they are responsible for maintaining the element of individual participation in social life by encouraging the feeling that it is possible to define ones work ones self, and thereby halting the characteristic imbalance in the population which is constituted by the numerous workers typical of large companies. However, it is at the same time, necessary not to become fixed on the dogmatic perpetuation of schemes which are incompatible with other constraints and opportunities which contribute to making a type of development sustainable. It is necessary, in particular, to exploit the skills generated by the existence of the university centre, by increasing the interaction between these institutions and all of the other public and private actors in social life. To consider the University in the historical centre merely as a generator of both greater presence in the city and of commuter movements is extremely limited. Instead, it must be thought of as an institution at the service of the social community since the University is able to make a formidable contribution in terms of innovation and of growth in both skills and capacity.

Respect for the Environment

The relationship between economic development and the environment is problematic in any place, but in Venice this subject has taken on dramatic proportions as a result of the Pandora's box which it is certainly not appropriate to open here. It is, however, useful to note that this Pandora's box is full of omissions, faults and failures as well as being the source of generalised and reprehensible misinformation (e.g., Frenna, 1972). Since there are no doubts that the time has come to turn the page, it is necessary to be clear about the limits and costs of eco-sustainable development or, rather, to make the probable and/or possible consequences of a certain planning decision known. Once this has taken place, let it be the social forces which determine its fate.

There are two clear examples of this. On the one hand, there is the coexistence of port operations with the hydro-geological equilibrium of the lagoon, whilst on the other, the continuation of chemical production activities and pollution. Social participation which is characteristic of the Agenda 21 process cannot be hoped for if, at the same time, information on subjects as complex and serious as these is only left to filter through the press or other unofficial organs.

This is a situation which must be quickly cleared up in order to make investment decisions. At the very root of this situation there lies the following fundamental question which must be posed in straightforward terms and, obviously, demands an answer: 'Will the Porto Marghera docks continue to be a place equipped for port operations, with the corresponding need to maintain the access channel beds at a certain depth and also with the consequences in terms of the hydro-geological, natural and biological transformation of the lagoon environment? Or, will these operations be renounced with the result that Venice will no longer be a Mediterranean commercial port?'

To this end, before making a definitive decision on the port channels, it is necessary to know:[11]

(1) the type of traffic (type of merchandise and goods transported) foreseen for the Venice Port in the medium- and long-term;
(2) the channel depths required, on the basis of the current and future naval technology, in order to grant the movement of the mix of traffic predicted, category by category (dry bulk, liquid bulk, container, unrefined oil, refined products, other cargoes);
(3) the consequences of the adoption of the canal depths required to maintain port operations in terms of the physical, biological and natural protection of the lagoon;
(4) the channel depths compatible with the physical protection of the lagoon and those compatible with its biological and natural protection;
(5) the foreseen consequences of adopting depths compatible with the physical protection of the lagoon as well as those compatible with its biological and natural protection in terms of port operations and the economy of the territory;
(6) should the two judgements prove incompatible, whether there are affordable technological solutions which would allow the problem to be solved (separation of the lagoon basins for example).

This analysis would be further complicated by the consideration of at least two more variables which are important both to port operations and to conservation. These are the high water events and the proposal for the temporary closure of the inlets on the one hand, and on the other, the tendency towards a rise in average sea-level in the long-term.

Should no answer be given to these macro-problems or other related problems through a reliable evaluation which can be understood by citizens, this will be tantamount to choosing neither to plan a development process nor to obtain social participation. It will, therefore, be tantamount to failing to achieve the objectives of Agenda 21.

Notes

1. Most of the information summarised here is from Randolfi, N., *Trasformazione urbana e produzione industriale nella Venezia dell'Ottocento*, in Var., *Venezia, città industriale*, Marsilio Editori, Venice, 1980.
2. For the brief information on the different production plants existing in Venice towards the end of the nineteenth century, the authors refer to 1887, as far as employment is concerned, since there is a specific publication (*Ingegneria a Venezia nell'ultimo ventennio*) which reproduces a report presented at the VI Engineers Congress and contains an estimation of the number of employed in the majority of plants existing that year.
3. It could prove interesting to develop an evaluation of employment in material production and of the production characteristics of island Venice after the first decade of the twentieth

century by using data collected by the Royal Central Institute of Statistics of those times, with the first industrial census carried out in 1911. However, since the aspect of the phenomenon's historical reconstruction is not a prime issue in this volume, but is only used as an introduction to evaluations of the current and future situation, possible in-depth analysis should be avoided here.

4. On these subjects, see also: *Marghera, rilancio a portata di mano* and *Porto Marghera cerca il rilancio*, in *Il Sole 24 Ore*, of 20.05.96 and 21.08.96, respectively.
5. See: Comune di Venezia (Venice Municipality), (1980).
6. The following constitutes a synthesis of a much wider and differentiated report; with the available space, it was not possible to show all the data and processing utilised to support the conclusions developed here.
7. See: Dardani, *La Fincantieri triplica i profitti* and *Fincantieri, il numero uno*, in *Il Sole 24 Ore*, of 03.04.96 and 16.09.96, respectively.
8. Statement of Ivano Beggio, President of the Venice Province Industrialists, in *La nuova*, 21, 03.11.96.
9. For an interesting comment on the critical decision-making issues, see De Rita (1993).
10. An interesting look at the recent past is found in Scaglione *et al*., (1993) and Scano (1985), p.369ff.
11. See also the introductory part of this chapter.

Bibliography

Amman, F.M. and Ninni, A. (1993) L'industria italiana della raffinazione: trasformazioni produttive, modifiche della domanda, sfida ambientale. *Economia delle Fonti di Energia e dell'Ambiente*, **2**, 159–193.

Benevolo, L. (Ed.) (1996) *Venezia. Il nuovo piano urbanistico*. Laterza, Rome-Bari.

Biondi, V. and Frey, M. (1995) Industria chimica a prova di Eco-management. *Impresa Ambiente*, **3**, 32–39.

Clò, A. (1995) Pianeta anno 2000: fame di risorse e voglia di sostenibilità. *Impresa Ambiente*, **2**, 6–11.

Comune di Venezia (1980) *Venezia città industriale. Gli insediamenti produttivi del 19° secolo*. Venice: Marsilio Editore.

Coses (1992) *Aggiornamento 1991. Indagini su Porto Marghera*, edited by Pugliese, T. report 128/1.

Coses (1995) *Porto Marghera. L'occupazione nelle principali aziende industriali di media e grande dimensione*, edited by Pugliese, T. doc. n. 729.

De Rita, G. (1993) *Una città speciale. Rapporto su Venezia*. Venice: Marsilio Editore.

Dekleva, J. (1972) I collegamenti tra l'isola e la terraferma. *Urbanistica*, **59–60**, 131–144.

Di Giulio, E. and Pizzoli, P. (1995) Some reflections on sustainability and its relationship with population. *Economia delle Fonti di Energia e dell'Ambiente*, **2**, 33–160.

Dorigo, W. (1973) *Una legge contro Venezia*, Roma, Officina Edizioni.

Ente della Zona Industriale di porto Marghera (1996) *Relazione anno* 1995.

Frenna, P. (1972) Inquinamenti della Laguna. *Urbanistica*, **59–60**, 21–130.

Frey, M. (1993) Strategie e strumenti gestionali nel rapporto impresa-ambiente. *Economia delle Fonti di Energia e dell'Ambiente*, **1**, 75–120.

Ganapini, W., Vianelli, L. and Zavatta, G. (1995) Via libera dall'industria lombarda alle tecnologie pulite. *Impresa Ambiente*, **1**, 47–55.

Grasso, G., Bufalo, G. and Manzo, G. (1994) La qualificazione ecologica dei prodotti industriali: il case-study della produzione conciaria. *Economia delle Fonti di Energia e dell'Ambiente*, **3**, 67–189.

Guelpa, F. (1993) Modelli di specializzazione e comportamento dei settori industriali. *Economia e Politica Industriale*, **80**, 51–175.

Malaman, R. (1995) Innovazione tecnologica e sviluppo sostenibile: le esperienze delle imprese italiane. *Economia delle Fonti di Energia e dell'Ambiente*, **3**, 173–201.

Malaman, R. (1996) Piccolo è bello, pulito è meglio. *Impresa Ambiente*, **6**, 34–43.
Mastino, G. and Radiconcini, L. (1996) L'Italia sostenibile. *Technology Review*, **89**, 68–73.
Pugliese, T. (1996) *La pianificazione urbanistica come strumento di politica industriale. La variante al Prg per Porto Marghera*, Series by the 'Istituto Nazionale di Urbanistica', Year II, n.9.
Randolfi, N. (1980) Trasformazione urbana e produzione industriale nella Venezia dell'Ottocento. In: Var. *Venezia, città industriale*. Venice: Marsilio Editore.
Scaglione, S. *et al.* (1993) *Venezia derubata. Idee e fatti di un ventennio 1973–1993*, supplement to n.20 of Avvenimenti.
Scano, L. (1985) *Venezia: terra e acqua*, Rome.
Unione Regionale delle CCIAA del Veneto (1996) *Relazione sulla situazione economica del Veneto*.

CHAPTER 6

Towards Sustainable Tourism in Venice

Jan van der Borg and Antonio Paolo Russo

The objective of this chapter within the scope of the volume is to correctly position the sustainable development of the tourism sector within the context of the Venetian society and economy and to provide the decision-makers with a support instrument, in the form of a coherent strategy for the development of tourism, in their resolution to balance growth and conservation.

Venice is known as a city of tourism *par excellence*. Its tourism generates consistent benefits by activating, both directly and indirectly, a vast range of economic activities and by permitting the local society to reach the critical mass necessary for maintaining social and cultural services which can only be found in larger cities. At the same time, tourism is considered a powerful constraint to economic and social development in that it leads to a slowing down of innovation and acts a trigger to the displacement of all activities with a capacity to pay less than the tourist industry. Furthermore, the environmental impact of the tourist activities can be serious.

Given the economic and social significance of tourism, to make its development sustainable becomes therefore one of the crucial points in any discussion on the city's prospects. This chapter is a contribution to the discussion in progress on policies for the management and reorganisation of tourism in terms of its sustainability.

In the first part of this chapter, a brief review is made of the principal themes which concern tourism in a city of art like Venice. In addition, the variables which determine the development of tourism in the lagoon city are highlighted, with particular reference to its environmental and socio-economic sustainability. In particular, various socio-economic indicators for tourism sustainability are used to obtain an acceptable quantification of the differential impact of the various sections of the tourist industry on the other sectors of the urban economy, as well as to identify some direct relationships between tourist and non-tourist sectors.

I. Musu (ed.), Sustainable Venice, 159–193.

This analysis, the most important results of which are presented in section 1.4, was conducted on a very detailed territorial level in order to thoroughly understand the qualitative differences between the areas affected by the large tourist flows on the one hand, and those which remain untouched by them on the other. The most significant conclusion is that the monoculture of tourism can not be sustainable in the long-term. The data shows how the economic pressure currently exerted by tourism, given the lack of an alternative production propensity in the historical centre, is changing the characteristics and the socio-economic functions of the city in an irremediable way.

In the second part of the chapter, two radically different scenarios for the development of tourism are outlined. Firstly, a no-change scenario in which the displacement of economic activities catering for residents by those for tourists becomes more acute, whilst the conflict over the use of the city between the resident population and economic operators bound to tourism becomes implacable. In the second scenario, linked to the presentation of management policies for tourist flows, some of these problems are resolved, even if not structurally. Tourism is developed in a way which is in synergy with the advanced tertiary sector, particularly information technology to support the cultural supply, which is reorganised according to criteria of financial autonomy, integrated management and value added production.

The analysis of that scenario for guided development, inscribed in the general framework of sustainability, leads the authors to some conclusive reflections about the potential strategic role of the city centre in the context of a metropolitan network with high levels of economic dynamism, such as the one emerging in the eastern Veneto or rather, in a broader context, the European urban network. An important precondition to being able to exploit such opportunities to the full is the organising capacity of the actors involved, as well as the quality of the required decision-making process.

1. Venetian tourism and the dimensions of its sustainability

1.1. Tourism in Venice, city of art par excellence

The enormous growth in demand for tourism which arose from the economic boom of the 1950s and 1960s has created a new market which is completely open to the discovery, creation and organisation of new destinations. Consequently, human and natural ecosystems have been modified to host the mass of tourists arriving. If, in recent years, the phenomenon of mass tourism has had a particularly powerful impact on the Alpine and coastal regions of the European continent, now it seems to involve, although with somewhat different characteristics, many urban destinations, especially those with a rich artistic and cultural heritage. This re-positioning of demand is not without consequences. Today, many cities of art live together with massive influxes of visitors and are already at their maximum capacity for receiving tourists.

Amongst these is Venice, whose tourist interest is firmly rooted in the culture and the traditions of this city.

An increasing number of visitors should have positive effects on the local society, particularly in terms of income and employment. However, it is logical to expect that if such growth continues unrelentlessly, the pressure of tourism on the historical city can become excessive: the number of visitors exceeds the carrying capacity for tourism, and the negative effects begin to prevail. Tourism can then cause damage to the environment, to the monuments, to the economy and to the quality of local life, and, in the last instance, also to the quality of the tourist experience itself.

The Venetian tourism product

We do not intend to describe at length the enormous wealth and variety which characterise the Venetian tourism product. Furthermore, there are few tourists coming to Venice who are drawn by the individual attractions rather than by the city as a whole and by the atmosphere which characterises it.

The majority of tourists visit the monuments which are grouped around the St. Mark's area. It is not surprising, therefore, that the routes which link the entry terminals with St. Mark's are often extremely congested. Moreover, there are numerous museums and galleries of great importance in Venice: the galleries of the Accademia, the Doge's Palace, the Peggy Guggenheim Museum, the Museum of the Fondazione Querini Stampalia, Ca' Rezzonico and the School of San Rocco. Palazzo Grassi, run by a giant private corporation, hosts a couple of exhibitions of international appeal every year. Events such as the Carnival, the Biennale, the Regata Storica and the Film Festival which are all held at fixed times of the year, attract a huge number of visitors to the historical centre. The main axes of the city and public transport are used by more than 20,000 people on peak days.

All this has had the consequence of triggering dramatic competition between tourists and residents over the use of Venice's historical centre which contributes in a decisive way to the intensification the exodus of families and firms. Mestre and Marghera, the urban agglomerations which are part of the municipality of Venice, are merely passive subjects within the local tourist system, little more than centres of accommodation which prosper from their easy accessibility to the historical centre on the one hand, and the national communication roads on the other. The Cavallino coastline, on the other hand, enjoys a seaside resort type of tourist attraction sustained, above all, by the camping component, which is independent of Venice as an attraction. However, the spillover effect which is produced by its proximity to the historical city is very high indeed, due to the phenomenon of indirect excursions, accounting for 30% of the tourists in the regional capital and is characterised by a very high spatial and temporal concentration. Therefore, together with the city's own particular physical structure, tourism is one of the main culprits of the chronic urban crisis in which the historical centre finds itself.

Tourists in Venice[1]

The residential tourism market has been in continuous expansion since 1963, in terms of both arrivals and presence. The total number of arrivals and overnight stays in the historical centre has risen from 758,975 and 1,731, 330 respectively in 1963 to 1,449,052 and 3,234,870 in 1995 (see Fig. 1). Except for some occasional lulls, growth was almost continuous until the second half of the 1980s. It is, however, quantitatively lower than that experienced by other cities of art. Once the existing accommodation supply has been saturated in the historical centre, the supply is slowly extended into the region and, at the same time, the phenomenon of day-tripping grows until it reaches dimensions which are much greater than those of overnight-stay tourism. The most significant transformations on the demand side in recent years are therefore the following: a marked growth in individual day-trippers and, more recently, in groups of day-trippers; a diversification in the visitors' motivations and types which take on the form of different segments of tourist demand; the day-trippers begin their trip from an increasingly extensive area; an increasingly large region for accommodating tourists. Therefore, day-trippers take on a fundamental importance. On the other hand, information on day-trippers is scarce and fragmented. Day-tripping does not involve a stay in accommodation. The visit to Venice is condensed into a single day. Every year, Venice is the destination for about 8 million day-trippers. The amount of day-trippers every year is about 80% of the total number of visitors (i.e. four day-trippers for every real tourist).

Three categories of day-tripping can be identified. First of all, there are the traditional day-trippers. They come to Venice from their place of residence. Indirect day-trippers (or day-trippers on the rebound) visit Venice via another

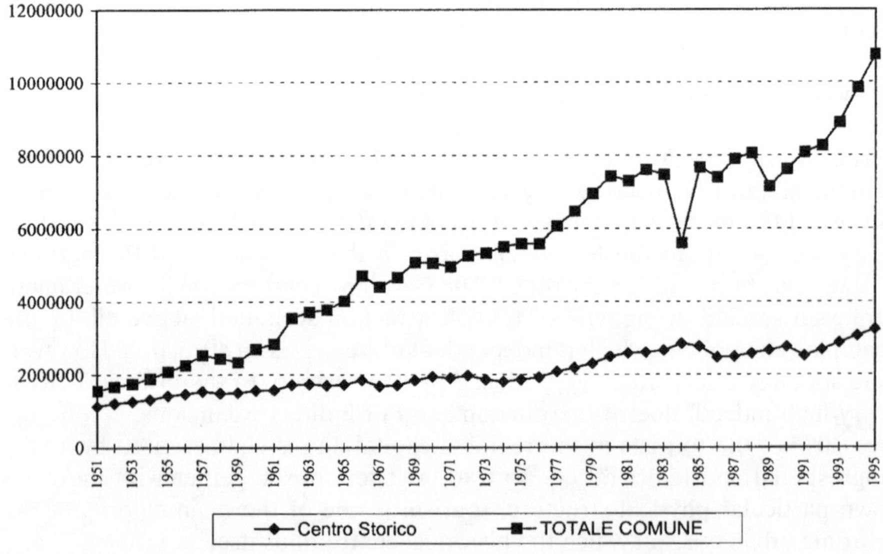

Figure 1 Overnight stays in hotels of Venice, 1951–1995.

place, which is the real destination of their journey; even the so-called passing day-trippers behave in this way, but they come from a place which is different from the one where they will then go to spend the night. The commuter day-trippers, or false day-trippers have chosen Venice as the destination of their trip but they stay in another place nearby, for economic motives. The number of day-trippers in comparison to the total number of visitors has tended to remain stable in recent years. The composition of these different categories of visitors is shown in Table 1. It should not be surprising, therefore, that the local tourist industry is embarking upon a process for re-orientating supply towards day-trippers. What is being witnessed is an expansion in the market aimed at day-tripping alongside a stagnation in the market aimed at residential tourism.

Rispoli and Van der Borg (1988) provide an explanation for the growth sustained by the day-tripper market. These day-trippers, who we have called false, save a great deal by staying in the periphery of the tourist region: hotel prices decrease constantly with the distance from Venice's historical centre, for the same hotel category. From this point of view, Padua seems to provide a very valid alternative to staying in Venice. A room in a four-star hotel in this city costs about a third of the price in Venice. The 40 km distance can easily be covered on Intercity trains in less than three-quarters of an hour.

Figure 2 shows the price gradient for hotels at different distances from the centre according to hotel category is shown. Such a gradient becomes increasingly steep as time passes. It is the enormous differences in price which explains the emergence of the false day-tripper, who wants to visit Venice but who spends the night in its environs. Particularly surprising is the fact that the prices of the hotels of the Lido of Venice are converging to the prices of those in Mestre; the Lido seems to have lost its appeal as a beach resort as such and is now becoming an appendix to the tourism of the historical centre.

The impact of tourism in Venice

Day-trippers spend less than residential tourists. In 1990, a tourist who stayed overnight spent on average 105,793 lira per day in the historical centre; traditional day-trippers spend 40,067 lira, indirect day-trippers 81,858 lira, whilst the false day-trippers spend as much as 104,062 lira, a large part of which is on accommodation outside the historical centre.

Table 1 Composition for category of visitors to Venice (August 1989–July 1990)

	Percentage
Over-night stay visitors	34.98
Day-trippers	65.05
Traditional	29.29
Indirect	11.31
Passing	5.51
False	22.90
Total	100.00

Source: Manente and Rizzi (1993)

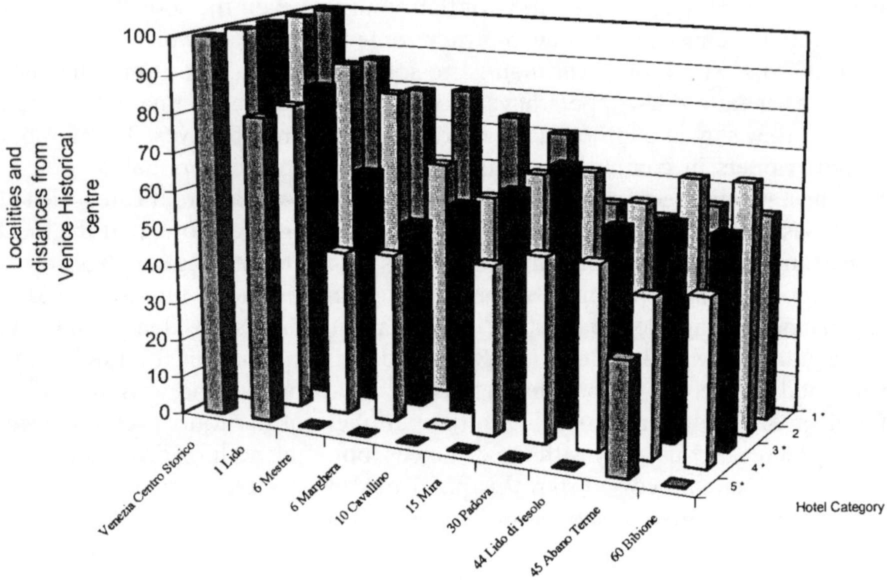

Figure 2 Price gradient for different hotel categories at different distances from the city centre (Venice = 100) 1996.

From the economic perspective, the preference for overnight-stay tourism is indisputable. Furthermore, the day-trippers' contribution to the various social costs is much greater than that caused by residential tourists. In other words, the economy earns more from tourists who stay overnight than it does from day-trippers. This contributes to widening the gap between the area enjoying the benefits deriving from tourism and the area bearing the costs and, hence, to determining a situation of structural fiscal crisis. That said, the case of Venice also demonstrates that the purely economic aspect cannot and must not be the only point of concern. It is, rather, the management aspect which makes day-tripping a serious threat to the city's tourist system. We will now attempt to explain this concept.

Residential tourism, by definition, depends on the availability of beds. This means that the maximum number of overnight-stay tourists which the historical centre of Venice can accommodate every day is strictly limited to the total availability of beds (slightly less than 12,000). Leaving aside the seasonal and short-term fluctuations, the price mechanism determines the demand for overnight-stay tourism. Furthermore, advance information on the demand of overnight-stay tourists is immediately available, since they tend to reserve the accommodation in advance. It is, therefore, relatively simple to avoid the inconvenience caused by 'peak' situations by using opportune regulatory actions.

Day-trippers, on the other hand, do not make use of any of the services which might lead to an absolute limit on their numbers. Regulation by means of price policies is thus very difficult to put in practice. The use of services which could in some way limit access to Venice, such as specific transport routes and parking places, cannot be made compulsory in any way. It is very difficult to predict their arrival and a systematic monitoring would lead to exceedingly high costs.

A further consequence of this model of the city's tourist consumption is that the city's enormous cultural supply is under-used (only 1 in 6 visitors who go to Venice finds the time to visit a museum of global importance such as the Accademia galleries) and visits are almost completely concentrated in the area around St. Mark's.

Therefore, controlling the flows in terms of sustainability requires tackling the problem of managing the day-trippers. Self-regulating mechanisms for this phenomenon do not exist in the same way as they do for residential tourism: appropriate policies must be undertaken to replace them.

1.2. The life-cycle of the cities of art and Venice's tourist carrying capacity

Two concepts borrowed from other economic disciplines help us to place the Venetian situation within a broader framework in which solutions in terms of management and organisational models can be looked for. The first relates to the cyclical nature of the phases which mark the tourist development of an urban destination and to the regulatory actions which makes it possible to determine its lasting success or its decline. The second is an attempt to obtain a numeric measure for the threshold of visitors beyond which the deteriorating effects of such a cycle manifest themselves.

The life-cycle of a city of art

The development process of every tourist site assumes cyclical features. The theory of the life-cycle of the tourist attractions is nothing but an application of the theory of product life-cycle used by market analysts in order to explain the fluctuations in the volume of the sales of a product.

The theory of the life-cycle of tourist places uses the number of visitors as an indicator. In its original formulation (Butler, 1980), the model of tourist life-cycle was formulated in absolute terms, i.e., in terms of the total number of visitors. According to this theory, every urban tourist destination experiences a starting phase, a taking-off phase, and one of maturity which can be followed by either a rise or by a decline according to the character of the tourism policies undertaken. An interesting extension of this model (Van der Borg, 1991) is given by introducing a qualitative element; the type of visitor. The study of the characteristics of tourism in the cities during different phases in their life-cycle indicates that not only the number of visitors varies, but also their mix, with decisive consequences in associated cost–benefit terms. As a result of life-cycle theory, what is commonly defined as the destination's carrying capacity can be compared with the long-term sustainability of its tourism development. Figure 3 illustrates this revised version of the life-cycle

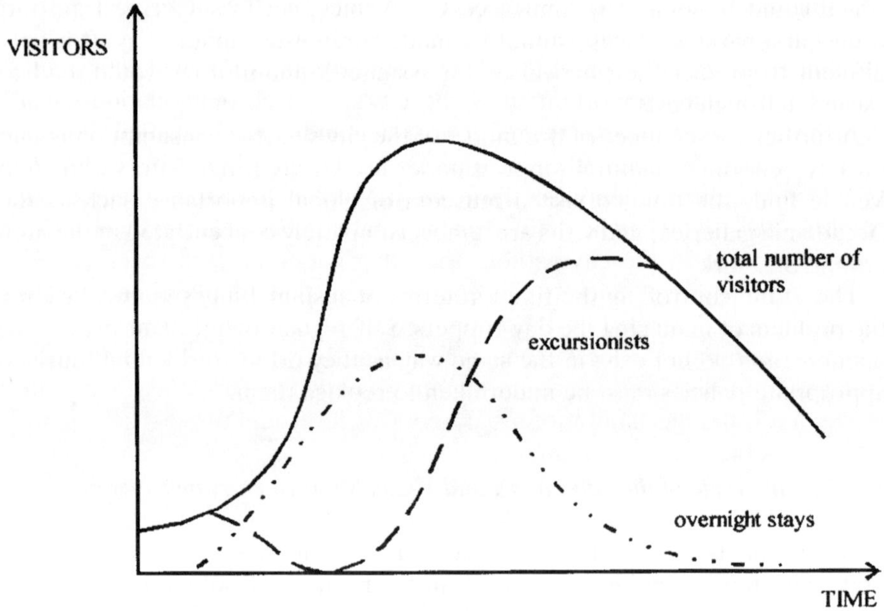

Figure 3 A revised tourist destination life-cycle.

of the tourist destination, incorporating the variations both in an attraction's absolute number of visitors and in their composition.

At the end of the cycle, in the absence of specific policies which might counteract the destination's decline, the city is transformed into a tourist monoculture and lacks every other economic activity which could compensate for the decline of the local tourist industry, with social consequences which can become significant.

It is difficult to avoid the dynamics described. The great pressure exerted by tourism in the final phases of such a process ends up consuming many of the resources at the city's disposal, some of which are not reproducible. The damage caused in this way to the image of the tourist city can also be irreparable. In addition, given that it is precisely the non-reproducible resources which play a key role in the initial phase of the cycle, it is reasonable to doubt that the city will be able to recover its position as a tourist attraction in the future. This point of no return is called the tourist carrying capacity of the city.

Venice's tourist carrying capacity

It is quite obvious that a city cannot accommodate an unlimited number of visitors: an upper limit to the tourist demand exacted on a destination or on a individual attraction exists, and this can be defined as the tourist carrying capacity. The life-cycle model for the tourist destination shown envisages the existence of a critical threshold in terms of tourist pressure, beyond which signs of stagnation and, in the end, of the decline in the urban destination's attractiveness begin to appear.

Numerous studies support the idea that the amount of individual attractions is not particularly important; it is the city's total accommodation capacity which determines to what degree tourist demand can be satisfied. For many cities, Venice included, the socio-economic carrying capacity is the most significant indicator amongst those mentioned in literature (Pearce, 1989; Onishi, 1994). This can be defined as the total number of visitors which can be accommodated in the city without the other economic and social functions being compromised. Such a concept is closely tied to the phenomenon of crowding out described by Prud'homme (1986). Tourism in the cities of art tends to dominate the local economy, eclipsing activities which are not directly related to tourism and which are driven towards the periphery of the urban system: it is the mechanism of urban rent which favours this displacement.

The problem of determining the socio-economic carrying capacity in the centre of Venice has been dealt with formally by Costa and Van der Borg (1988). Such model interprets the conflict taking place between tourists and resident population over the use of the urban functions.[2] It is based on the maximisation of the income produced by tourism within certain constraints related, for instance, to the availability of accommodation, refreshment services, car-parks, urban transport, the capacity for rubbish disposal and the space available in St. Mark's Basilica. Each destination will then have a carrying capacity determined via the consideration of its core functions.

With reference to the estimates made by Costa and Van der Borg (1988), the critical limit of about 25,000 visitors a day[3] (of which about 14,000 are day-trippers) was exceeded on 156 days in 1987 and will be exceeded *ceteris paribus* on 216 days in the year 2000.[4]

Two other indicators which reflect the inconvenience caused by an excessive tourist pressure in a significant manner are those which compare the number of visitors and number of available beds with the resident population (cf. Van der Borg and Gotti, 1995). The trend in the tourist pressure index calculated for the different areas of which the municipality of Venice is made up is illustrated in Figure 4. In Venice, there are about 50 tourist overnight stays a year for every resident in its municipality, in comparison to 36 in Salzburg, 23 in Bruges, 10 in Florence and 6 in Amsterdam, other mature urban tourist destinations which compete with Venice in some way. This number rises to 175 if the population of the historical centre alone is considered. As far as the index for tourist intensity is concerned, the number of beds per inhabitant reaches 15 in the historical centre and 7.3 in the municipality of Venice, in comparison to more or less analogous figures for Florence, 7.4 in Salzburg and 6.7 in Bruges.

A fact emerges from the comparison of these indices: whilst the accommodation supply is relatively equal in all of these places, Venice has to deal with a vastly larger number of visitors. This contributes to exacerbating the congestion problems which in all of the other similar places appear in much

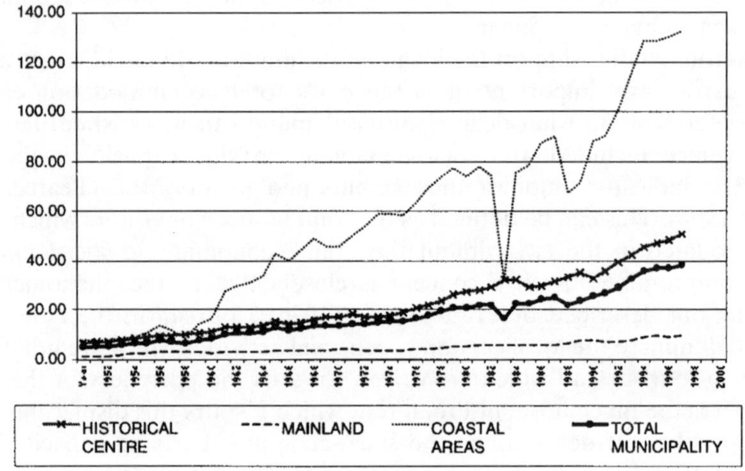

Figure 4 Tourist pressure ratio, different areas of Venice municipality, 1951–1995.

milder forms and can be dealt with without being placed in an emergency context.

The situation in Venice is so much more dramatic if it is considered that in the other cities of art examined, management policies for the tourist flows have been implemented for as long as 15 years. These have already mitigated the most serious problems connected to congestion (decentralised car parks, diversification in city access and in the means of transport, widespread attractions, etc.).

1.3. Some measurements of the sustainability through indicators

These aspects related to the pressure of tourism on the city and to its life-cycle position are, in themselves, sufficient for establishing a picture of tourism's complete unsustainability in Venice. This analysis can be sub-divided in order to examine other aspects of sustainability, such as the physical and economic ones, or, rather, to investigate in greater depth the socio-economic determinants which are summarised schematically by a high ratio of visitors to residents and by a low ratio of tourists to day-trippers. Various indicators have been developed which are suitable for illustrating the situation of Venetian tourism.[5] The most interesting results derived from the use of such indicators concern the following subjects:

- *Tourist use of public transport and of private transport*. This accounts for 10% of passengers on urban bus services and for 18% on extra-urban routes; for about 20% on water routes, with a concentration which causes disruption in peak periods partially overlapping with resident traffic rush hours. More than 50% of boats registered with the Harbour Master's Office in the last 15 years are for tourist use.

- *The collection of waste produced by non-residents*. The amount of waste collected that is produced by overnight-stay tourists is about 10% of the total (from 1% in 1980); the percentage of waste produced by day-trippers is estimated at about 83% of the average annual total, on the increase and concentrated during summer months (see Fig. 5).
- *The socio-economic impact of tourist pressure (indices of specialisation and of concentration)*. About 35% of the economic activities in the historical centre is made up of commercial activities for tourists; employees in the tourist industry account for 19% of the historical centre's population, while the local units of a tourist nature are 23% of the total. All of these figures have been increasing over the last 5 years.
- *Tourist prices*. The price of a hotel room in a four star hotel in the historical centre of Venice is triple that of a hotel of the same category 35 km away (Padua) and double that of one 6 km away (Mestre); Italian tourist prices rise more than those of all the European countries which generate tourism and only increase less than prices in Greece as far as countries competing in the tourist market are concerned.
- *Accessibility of the Venetian cultural supply*. The indices of opening for Venetian museums (opening hours out of total annual hours) range from 12.7% for the Natural History Museum to 40% for the Doge's Palace; the indices of availability (m^2 per visitor/opening hours) range from 33 m^2/hour

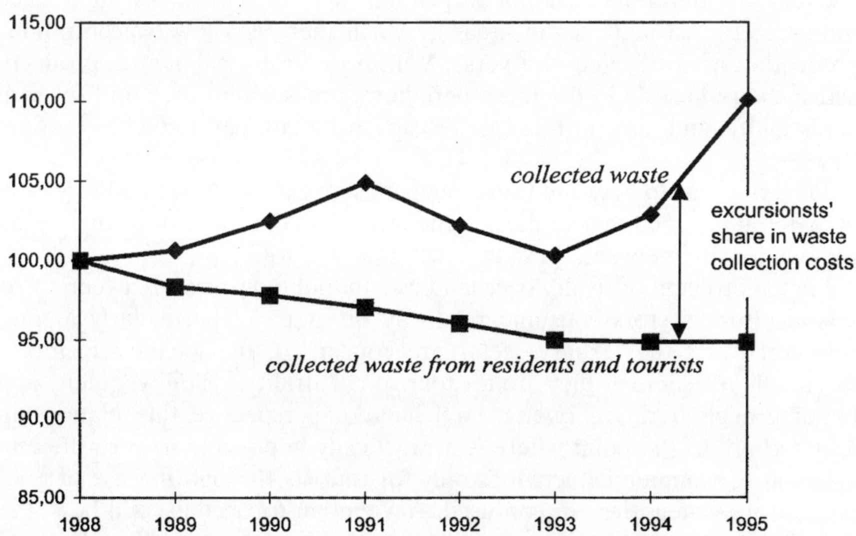

Figure 5 Waste collection, values per visitor (upper line) and per person sleeping in Venice (lower line). Years 1988–1995, Index numbers, 1988 = 100.

for the Doge's Palace to 360 m^2/hour for the Correr Museum (Funari and Simionato, 1996).

- *Qualitative standards of complementary supply.* In 1994, 65% of hotel rooms were not equipped with air-conditioning, 70% did not have a telephone, 80% did not have a television set, 68% did not have services for the disabled, and only 9% had the capacity to host conferences. Yet, the room occupation rate is close to 100%.

The picture which emerges from these values (and, where available, from their associated trends) is alarming and necessitates an analytical examination in terms of the conflict with other economic sectors over the use of urban space.

The importance of such values lies in the possible use of them as a monitoring system, should the local government want to consider policies aimed at bringing the system back on a course towards sustainability. Actually, such a course was embarked upon by the city of Venice with the start of the Local Agenda 21 project in accordance with the principles of the Aalborg Charter, which obliges the member municipalities to establish a participative process for the creation of a common vision and of shared policies towards sustainability.

1.4. Tourism and other sectors of the local economy

The interrelations between the material production sector and the tourist sector seem, at first sight, to be very weak; real and proper competition does not exist as far as land use is concerned, and the two sectors' location factors are radically different. It is not surprising, therefore, to find a high concentration of tourist activities in areas in which there is a low concentration of production activities and vice versa. Within the historical centre, production activities are located in the more peripheral areas where they find space and accessibility, and also, in this case, escape from competition with the tourist flows.

The relationship between tourist activities on the one hand and service and immaterial production activities on the other proves more interesting because the relationship between these two sectors is two-fold.

First, a problem of rivalry over land use undoubtedly exists between the two sectors. Both sectors consume relatively little space; particularly as far as commerce and the tertiary sector are concerned, the location factors are essentially the same as they are for tourism: centrality, visibility, quality of the building, agglomeration. Even from the market perspective, this relationship is easy to find, to the point where is it practically impossible to curb the establishment of commercial activities only for tourists through the use of grants, because the categories are confused. A Venetian tobacconist will be unlikely only to sell goods for smokers; from here, it is a short step to selling almost only souvenirs. For service and research activities, the matter is slightly different. Increasingly often, a transfer of these institutions to abandoned or marginal areas of the historical centre is witnessed. Here, they find a space of adequate quality and efficient reorganisation. Indeed, even the university and several other significant public institutions are relocating to these areas while their original sites slowly lose their significance and strategic advantage.

Obviously, however, there are also points of synergy between the tourist industry and the other services. These are particularly plentiful and promising where residential and cultural tourism is concerned, considering the complexity of the production chain of these two sectors. For several of such activities, and especially for those with high value added such as, for example, research centres and centres producing software and other hi-tech services for businesses, the physical proximity of the primary activities with the service ones is not so important. What is a key factor is the circulation of knowledge, the setting up of common infrastructure, strategies for integrated development. For lower order services with less value added, such as retail trade, art galleries, exchange bureaux, photocopying shops and university bars, the physical proximity becomes the discriminating factor for success in the relationship between cultural attractions and tertiary sector enterprises.

In the historical centre of Venice, some immaterial production sectors are very developed, in particular the retail trade. It is no coincidence that the great economic tradition of Venice was that of trade capital. Until a few years ago, the Venetian population could rely on a commercial per capita revenue which was even double that in many hinterland areas. The explanation for this phenomenon's survival lies in demand in terms of an absolutely exceptional equivalent population. So, the phenomena of discontinuance and replacement which have been observed recently, as the consumption habits of Venetians change and part of their demand is shifted onto the hyper-markets in the periphery, are to be considered natural; on the other hand, the architectural and logistic constraints do not allow the establishment of large retail distribution services in the historical centre. It is, however, the elements of quality and welfare which make Venice's situation become worrying. Indeed, the retail trade can rely on very high margins originated by a tourist demand which is highly inelastic with respect to quality because many do not return. This and other characteristics of the demand, such as the limited space in the city centre, push consumption prices to a much higher level than can be explained by the cost differentials of transport and storage of goods. Since the tourist demand is made on the same concerns which are used by residents, the latter sustain a heavy loss in terms of well-being.

That said, the districts affected the most by overnight-stay tourism and those which are most directly influenced by the tourist flows are also those which show the highest concentration of business catering for residents, whilst some areas which are less affected by mass tourism are those which are more lacking in such provision. The sectors most drawn by the presence of the tourist industry in the same area are those of banks and of services for businesses; education, health and public administration tend to be located in a way which bears no relation to tourist flows. Business services, or at least that part of them which forms a more advanced sector from the technological point of view and which provides these businesses with the highest level of value added, are not necessarily located in physical proximity to cultural attractions and hotels which represent a significant portion of demand for these services. Indeed, they are located all over the territory. Finally, the other services (amongst which non-tourist transport, social clubs, sport, etc.) show a fairly marked inverse correlation with the presence of tourist activities.

The areas which show a more distinct vocation for tourism are those in which a higher percentage of unoccupied houses is found, with a maximum of 9% in the area around the district. On the mainland, this percentage drops to roughly 3% and reaches the municipality mean value in the only area affected by the overnight-stay tourist flow. The correlation with the percentage of houses which are for sale or for rent is not so marked. If, on the one hand, the tourist market ought to lead the apartment market, on the other, a certain saturation is seen in this market, which remains quite dynamic in the areas where land values are higher.

Moving on to consider how the nature of tourism in an area affects its demographic composition, the fact that the historical centre is becoming increasingly depopulated of young people is already well known, and this phenomenon already affects the more central districts of the mainland and the Lido as well. In the same areas, the percentage of residents over 65 years old is high and these residents are particularly numerous where the economic base is more bound to activities of a tourist nature. Within the historical centre, the inverse relationship between the residents' age and the degree of tourism in the area is also evident. However, there appear to be fewer elderly people in areas where the replacement of basic services by tourist activities proves more marked. These are the same areas in which the highest rate of depopulation has occurred over the last few years. In these few figures, the whole trend of the exodus which affects the historical centre can be read. The pressure of tourism makes the land value increase, pushing out those residents who cannot afford the high rental or maintenance costs as well as preventing new families with children from moving in. The scarce demand for primary goods accelerates the process of replacing activities for Venetians with activities typically for tourists. This replacement process is less evident in districts which are less affected by tourism. In these areas, the insufficient demand leads to the progressive closing of shops and of services for the population and to the driving away of artisan activities, but these are not necessarily replaced by tourist type activities. In these areas, the urban values are particularly low and the elderly continue to live there, in increasingly worse conditions in terms of welfare.

1.5. Some Conclusions. Should we worry?

We shall now try to bring together the threads of this analysis, focusing on the more peculiar dynamics which affect the areas with most tourism in the municipality of Venice, that is, the historical centre, the island and coastal areas. In fact, the data for the mainland shows a make-up and a socio-economic growth which does not particularly diverge from the other centres in the Italian north-east. The economic trends, especially in the tertiary sector, are fully integrated into the qualitative restructuring which is taking place in the industry of the Veneto. The crisis of the industrial centre of Porto Marghera does not imply future social damage since it is a phenomenon which has already been consolidated and which, at least until now, has been reabsorbed by the local economy without any trauma. In addition, the influence of tourism on the area is beginning to become significant. However, the mainland does not have to sustain the load of social costs and external diseconomies which burdens the historical centre. Congestion is non-existent, tourism is one of the activities, experienced indirectly, which feed the local economy and the principal route to economic development certainly lies more in the projection of production and of business in Europe than in the creation of structures for tourists or museum networks.

In contrast, where the other areas facing much more obscure fates are concerned, on the basis of the analysis carried out, we can identify several areas which have homogeneous features.

Along the main itineraries, the make up of the supply for tourism is dominated by small commercial activities catering for tourists and, in particular, for day trippers who are easy to please. In these areas, many medium category hotels and the majority of the restaurants are also found. In the central and most congestioned areas, which include St. Mark's, the highest concentration of cultural and museum attractions is also found. Around these and in the most interesting places in terms of scenery, there are many street vendors with both permanent and mobile positions. In the same areas, the mortality rate for concerns typically catering for residents is high, primarily for the food trade and for artisan activities which, more often than not, are replaced by activities which cater for tourists. The urban values and the prices of consumer goods in these areas are much higher and, as well as the high prices, the resident population also suffers from aesthetic degradation and the congestion in pedestrian traffic and in transport. Not even the tertiary sector enjoys good health. Indeed, the transfer to the mainland of the headquarters of banking, publishing, and insurance companies is witnessed. The creation of new service activities and notary offices stagnates, while a certain dynamism affects the clothing sector.

The areas just outside the centre are those which, although not directly affected by the large flow of visitors, benefit to a greater extent from the overnight-stay kind of tourism. Here, the replacement of activities catering for residents with tourist ones does not go beyond a natural level, depopulation is not as high and tourism does not create serious inconvenience for the resident population, which can continue to count on good commercial supply and on a good quality place of residence. The penetration of the trade of trivial products for tourists has not yet reached worrying levels, even if there have been signs of a strong increase in this over the last few years.

In the marginal areas a large flow of visitors rarely arrives, except for the few 'cultural' tourists who go on foot to visit the Biennale, located in a slightly marginal location, or those who follow the museum itinerary Palazzo Grassi-Accademia-Guggenheim in search of peace and artistic charm. In any case, while the autonomous demand generated by the area is not in a position to sustain the existing supply which is gradually reduced, new concerns and services are rarely located in these areas, despite the low urban values, because of the inaccessibility which characterises them. These take on the character of areas of economic desertification. Some recent decisions made by the municipal administration such, as the *PRG* (town-planning scheme) for the historical centre and the *Piano per l'Adeguamento della Rete di Vendita* (plan for the adaptation of the sales network) could constitute redeeming opportunities for these areas. The new town-planning structure for the Giudecca island, the recovery of an area of great strategic importance such as the Arsenal, the creation of a permanent museum centre in the Biennale area in Castello, the establishment of the Ca' Foscari University at the city gates close to the railway station: each of these ventures constitutes an opportunity for sustaining demand by making them accessible to students and tourists, and thereby allowing the maintenance of a certain economic structure or the revitalisation of tertiary activities. The university areas already enjoy their own dynamics and seem to be the only ones whose economy is not dependent on the tourist economy.

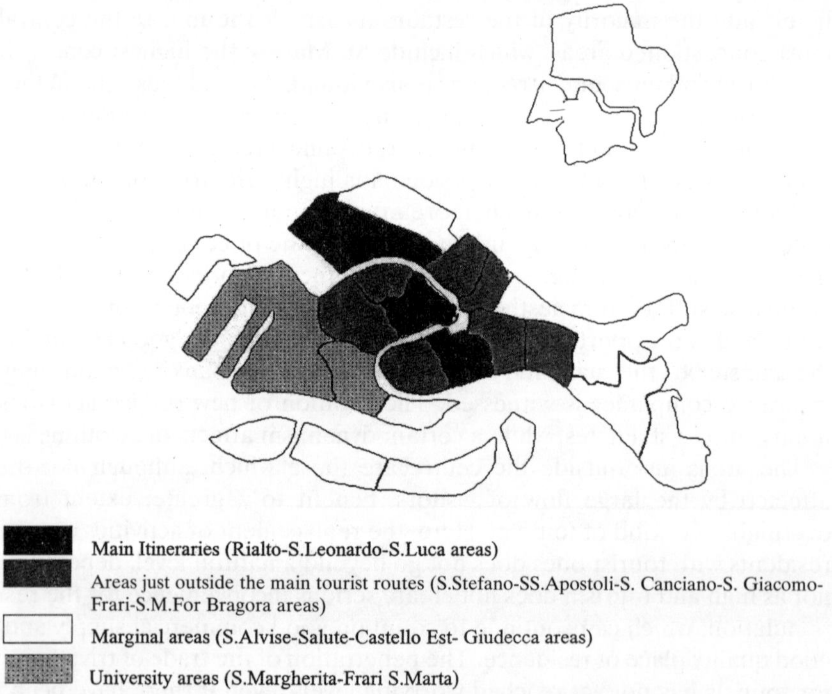

Figure 6 Zoning of Venice according to the relevance of visitors' flows.

Finally, the island of Murano, whilst being a decentralised area, has remained Venice's only real industrial centre and it continues to generate wealth and employment with a market power (in terms of international export) which is sufficiently independent of the local tourist market. Furthermore, it is one of the few islands able to count on quite strong and regular connections with the historical centre. Even Murano, however, loses commercial activities for residents while the artisan glass production activities grow consistently.

As far as the coastal area is concerned, a marginal area can be identified in the island of Pellestrina which lives almost exclusively on fishing and its dockyard and is affected by the tourist flows in a totally negligible way; the Lido is a well-served area and maintains a certain economic and commercial structure, while the area which undoubtedly has the most interesting development is the Cavallino, the only area within the municipality of Venice to grow in demographic and economic terms as a result of the entrepreneurial characteristics of the existing supply.

Taking this map of Venetian tourism thus drawn into consideration, we intend to set out an urban policy strategy which aims to contain the negative effects of the tourist flows in the areas most damaged by them and to spread the benefits to the areas which presently do not enjoy them in any way. This must inevitably be accompanied by a policy for regulating the tourist flows, by raising the quality of the hotel supply and an increase in hotel (and extra-hotel) capacity catering for segments of demand which, currently, are poorly supplied, as well as by the redirection of part of the tourist demand towards cultural tourism with the use of appropriate marketing action.

However, it is clear that all this might not be enough should tourism continue to remain the only economic activity for Venice's historical centre. For this reason, the strategic aspect regarding the policies on tourism is inexorably bound to the fates of the immaterial economy and of services and to a drastically innovative transport policy.

2. Venetian economic development scenarios linked to tourism

In the preceding paragraph, some distinctive features of the model for the tourist industry's territorial development were outlined.

The principal fact is that it is not or, rather, it is no longer, possible to divide the historical centre into zones according to economic vocations, as has emerged from previous research (cf. Lando, 1986), in which a typically commercial–food area could be identified, as could a CBD with a high concentration of banks, private offices, high quality trade, a notably residential area. The demographic drop and the exodus leave scorched earth all over the Venetian territory, and tourism does nothing but take what remains for itself, with concerns of the lowest value added but which are profitable enough to guarantee them a certain economic life. The commercial area of Rialto yields an increasing amount of space to souvenir stalls. It is the same proprietors who, faced with the already desperate situation of the commercial and distribution logistics which affects the retail trade, prefer to give up their activities and sell to whoever can pay. The business area remains affected remarkably little by the large flows, and maintains a certain qualified economic and residential propensity. But for how much longer? Some time ago, the closure of a historic food shop caused a great uproar: the event was interpreted by many authoritative commentators as the sign of the city's definitive yielding to tourism even in its more intact areas, or, rather, in its most distinctive features of Venetian identity.

There is no point in emphasising how, along with the shops and banks, (at this point its seems even desirable) the Venetians leave as well. Above all, however, firms and jobs disappear. Indeed, what advantage can any kind of firm have from being established in the centre of Venice and thereby taking on enormous costs for the purchase and maintenance of space, enormous accessibility and mobility problems, not to mention emergency situations such as high water, only to be confronted by a local demand which is practically non-existent in relation to the rich and dynamic demand which characterises the urban agglomeration of Padua Venice and Treviso?

Therefore, in terms of physical environmental sustainability, tourism is certainly not the most damaging industry for the fragile lagoon ecosystem and, indeed, requires an environment which is not degraded in order to prosper. However, in terms of the socio-economic sustainability, any course for developing the Venetian economy which did not succeed in somehow slowing down the pressure exerted by the tourist industry on the local economy appears a lost cause from the start. Once it has been accepted that the maintenance of Venice's function as a city is a quality to preserve, the data for the exodus, for the trends in the demographic and socio-economic structure, and for the environmental quality itself, indicate the serious risk of tourism remaining the only economic activity in Venice. Moreover, as has been seen, it is an activity which, by its very nature, is exposed to shocks and subject to the danger of self-destruction. Other sectors will simply give up space to this force and will relocate in the thriving and dynamic economic ground provided by the rest of the region. More sophisticated analyses can indicate which mechanisms bring about this process of impoverishment and of replacing the local economic base, but some statistics are sufficient for obtaining a picture of the force of the phenomenon in progress. With a tourist industry which by now, including only those firms directly involved in tourist production, accounts for 25–30% of the firms and of those employed in the historical centre; with material production in decline in terms of employees and firms, even leaving the situation of Porto Marghera out of consideration; and with the figures for immaterial production, which ought to be the sector to rely on, but which is essentially stagnant as compared to the 10–20% increase in this sector in other cities of the Veneto. Even Venice's administrative function, the only one not to have declined in the last decade along with education and health, could be doubted in a prospective redistribution of local powers based on very real considerations of competence which the Italian political system is increasingly approaching.

This pessimistic but highly realistic scenario must be overturned using the few weapons which remain available to those who have the city's future at heart. This must begin from the city's only true vocation, its powers of attraction for tourist and representative purposes, i.e. its beauty, which must be sold at a high price and made profitable in harmony with other sectors rather than to their detriment. Cultural tourism and cultural production in general are expanding areas but ones which have not yet exploited their potential to the full. Museums, run in a strategic and integrated way, will have to become financially independent centres of cultural production able to attract a much larger public than the present one. If just one out of every three visitors who walk every day along the streets of Venice went to one of its thousand cultural institutions and spent part of their budget there, rather than in souvenir shops of dubious taste and authenticity, that would enable an authentic and effective financial management of the Venetian tourist system. The first consequence would be that many problems of channelling resources for financing the big projects needed to preserve the attracting power of the city in the long term, not only with respect to tourism but to all sorts of strategic activities, would not exist.

The cultural production, which involves universities, research centres and the historical–architectural heritage, is a strategic element in Venice's development. As well as being perfectly in line with the city's millennial history, cultural production becomes an element of the framework of the Italian north-eastern economy, thereby permitting Venice to overcome its marginality in geographic terms and, at the same time, providing the whole area with logistic and directional support which is of high quality and unbeatably attractive. This does not make it possible to resolve the incompatibilities between tourism and other sectors in structural terms, but it does make it possible to render the environmental quality profitable, to give the existing cultural structures a self-propelling capacity, and to stimulate growth in the immaterial sector.

Instead, what we are witnessing is a dramatic phase in the dichotomy between benefits and negative externalities (physical congestion, trivial products, urban degradation, small-scale crime) imposed by the attractions on the well being of citizens and on the very quality of the experience enjoyed by visitors. The city universities, the only assets still in a stage of expansion, take possession of every free space, giving the impression of colonising the historical centre and of being the primary cause of the exodus. However, they do not succeed but to a minimal extent in passing on education and knowledge to the Venetian economic environment. They will have to become the real drivers of development by integrating themselves into the social fabric and providing an educational and research output which can be really invested in the city.

Although the problems related to the pressure of tourist flows are perhaps more serious in Venice than in any other European city of art, a real strategic planning for tourism development has never been carried out, even though various feasibility studies on policies for controlling demand have been initiated. However, to ease the pressure of tourism on a site, the supply can and must be acted upon, expanding it as much as possible and encouraging visitors to try alternative, non-central routes by means of a suitable policy of information on and commercialisation of the product. In other words, it is necessary to break the vicious circle which stimulates a supply of increasingly trivial tourist products and to encourage, instead, a virtuous circle which begins with the identification of areas with an unsatisfied need for the tourist clientele, in order to create quality, selective and high value added tourism.

An improved accessibility, not only physical; the full integration of the universities and of the other knowledge production centres with the production structures and the subsequent contribution of knowledge, instruments and strategic vision to the tourist and cultural industry and to urban management; the widespread establishment of advanced services for businesses which can lend value to the hundreds of city cultural institutions and to events producers; the planning of visits, made possible by information technology; the comprehensive promotion of the city and of its events; the improvement in the quality of residence through appropriate town-planning and fiscal instruments; the political representation of the interests of all the users of the city, also, therefore, of those who work, live and spend most of their lives in the city whilst being resident elsewhere; all these are elements in a comprehensive strategy for revitalising the Venetian socio-economic fabric which envisage the tourist sector as being intimately tied to the immaterial production sector, within a synergetic development which exploits the tourist vocation in order to guarantee an independent growth beyond tourism but which is fully within the economic system of the north-east. This seems to be the only route to take in order to save Venice from an inevitable destiny of plastic gondolas and long queues at the water-bus stops.

The effectiveness or ineffectiveness of this strategy can determine, in the medium-term, the occurrence of two scenarios, polarised for simplicity:

(1) The no change scenario, which is distinguished by an uncontrollable growth in mass tourism, with a persistent worsening in territorial and functional imbalances in the tourist system; the reduction in the quality of tourism understood as the 'cultural content' of the visit; the further diffusion of commercial activities bound mainly to flows of day-trippers. In other words, the city would be in the declining phase of the curve of the tourist destination's life-cycle. Furthermore, the points of incompatibility between the city's tourism development and economic development and, therefore, the conflict between the population dependent on tourism and the rest, would escalate to the point of causing definitive structural changes in the economic composition of the city. Evidently, this scenario is unsustainable because the city, lacking an alternative production specialisation, can neither get rid of tourism, nor, for the reasons seen, can tourism alone maintain forever a profitability which is sufficient for guaranteeing the city's economic vitality. In any case, it is held that an economic monoculture, whatever kind it might be, exposes the local economy to shocks which are unlikely to be endured in the long-term.

(2) In contrast, there is a scenario for the virtuous development of tourism which could be based on specific actions within the scope of an integrated urban strategy; these actions ought to give priority to cultural tourism, favouring the internalisation of benefits generated by the flows of day-trippers, the increase in the quality and the value added in the services supplied, the reorganisation of cultural production, the territorial and seasonal decentralisation of the flows, the stimulation of latent demand sectors, the application of telecommunications and information technology as further instruments for forecasting and for regulating the flows.

Between these two extremes, there is a whole range of intermediate actions which are still essential for curbing the corroding effects on Venice's physical and social environment caused by the present model of tourist exploitation. We have intentionally excluded a scenario in which tourism, as a whole, takes a considerable step backwards. Indeed, Venice still maintains an exogenous and constant attraction for tourist purposes within a world-wide tourist trend in constant growth. To consider closing Venice to tourism means excluding it from its only real production vocation which currently exists (Russo, 1998). For this reason, we maintain that to drive tourism away by closing the bridges and opposing the peaceful invasion which affects Venice every day, bringing prestige and prosperity to the city, is, therefore, a counterproductive and senseless policy.

The contrasting model, that of a total abandonment of the city to its fate as a tourist attraction, the creation of the open-air museum feared and opposed by many, on the one hand, appears to be an unavoidable fate, beyond any remedial action, if the city does not find itself an alternative production vocation to the tourist one. On the other hand it is an asymptote, and as such impossible to obtain without provoking the city's social death at the same time. As has been shown above, Venice is attractive because, as well as being beautiful, it is also a city, with its inhabitants, its traditions, its rhythms of life and its rich cultural heritage, which includes the immaterial heritage embodied in citizens and functions.

The objectives of a tourism policy which could be directed towards sustainability, within a context of a production transformation in Venice's historical centre based on finding a vocation as a technological research centre with a high added value which could further distinguish Venetian tertiary production, can be summarised as follows:

- the containment of congestion caused by the flows of visitors to Venice, by means of a redistribution of the peaks over the whole year and the lengthening of the season, in such a way as to keep within the carrying capacity's maximum limit (fixed at around 25,000 visitors daily) for as many days as possible;
- a diversification in access terminals and in itineraries within the historical centre;
- the historical centre's closer connection (both in a material and in an immaterial sense) with the mainland and with the lagoon environment, which must be 'opened' to tourism without disturbing its environmental features:

- more generally, curtailing the loss of welfare experienced by the population as a result of the economic pressure exerted by the tourist activities;
- an increase in the quality and the transformation of the very nature of the museum production process, so as to fully support the city's cultural vocation by triggering the economic dynamics of this high value added industry;
- a political objective of equity in the tourist sector, or, rather, the prevention of fiscal evasion and of unlawful activities.

These are the differing scenarios which are anticipated in relation to the choices on Venetian tourist development. It remains to be seen how far the millennium year of 2000 will provide the first opportunity, and one which is already imminent, to see which of the policies should be implemented. In that year, an enormous flow of visitors will enter Italy and Venice will be heavily affected by it. The ALATA system has been designed to manage and distribute the visitors' flow in the whole Northern Adriatic area in such a way as to offer a balanced and comfortable network of opportunities for those who come to Venice during the peak months. If it works efficiently, that will be a good start for a long-term strategic planning for tourism on a regional level aiming at making all the actors involved better off, visitors, citizens and tourism operators, and it could represent a stimulus for triggering a virtuous circle of tourism. On the other hand, a failure or a delay in the implementation of this system would mean for the city risks total paralysis, and an extremely high price to pay.

2.1. Regulating the flows

An urban policy measure which aims at managing the tourist flows which come to the city of Venice can be carried out by applying processes of a soft nature, adequately planned and assessed in terms of their effectiveness. These instruments are preferred to hard measures (closure of the urban centres, restrictions in car parks, prohibitions, etc.) not least because the other European cities of art which have tried to resort to them (e.g., Salzburg, Toledo, Oxford) have had to give them up within a few years, since the market forces were such that these constraints were avoided and made useless or too burdensome. It is presumed that such processes could modify the visitors' behaviour and modify the actual composition of the demand segments in order to make it more compatible with the existing cultural structures (and with those which are being realised on the basis of the municipal plan for reorganising the museum system). It is aimed to bring about the transformation, by means of an effective and all-inclusive promotion and the extensive diffusion of up-to-date information, of a markedly day-tripping based tourism into a tourism which is of a more permanent kind. Included in this aim is the prolonging of the average length of stay of the present, more permanent tourism and the circulation of a larger number of visitors both within the historical centre and in the immediate lagoon surroundings.

The utilisation of information technology can facilitate the identification and the management of the different demand components, even only potential ones, which come to Venice as well as the provision of a suitable cultural and complementary supply. To prioritise accommodation bookings (and, possibly, the booking of parking space at the terminal), together with all or part of the tourist package, is a fundamental part of this policy because it makes it possible to provide a quality tourist product through anticipating the peak periods in which congestion and inconvenience for the citizens can take place. At the same time, this will act as a disincentive to the spontaneous visits of the day-trippers.

In addition, this strategy foresees the issue of a multi-service card; as a result of an opportune marketing strategy, the visitors will be invited to book their trip to Venice, and they will receive in exchange a Venice Card which offers them a series of concessions and opportunities which are not available to those who do not book (still granting the latter free access to the city). The number of cards issued will be equal to the tolerance threshold which is periodically determined on the basis of values of parameters related to the correct functioning of the urban system. The card will be sent, at no charge, to overnight-stay tourists when they make the booking for their hotel room, so as to encourage optimal behaviour of the visitors to Venice, in terms of times, activities, seasons, etc. These tickets would be a system of payment for entrance into Venice which is both juridicially possible and socially acceptable and which, in association with tele-booking, would produce an intelligent system of limited numbers and of spatial and temporal decentralisation of the peaks (Di Monte and Scaramuzzi, 1997). In this way, the decision to sleep in the historical centre, thereby taking advantage of the above-mentioned benefits but being forced to book, must be compared with the relative flexibility of not booking the visit and staying overnight where it proves more convenient (which, therefore, risks congestion).

A first step in this direction will soon be made by the municipality of Venice which intends to tender to a private partner the electronic ticket-office with tele-booking for the civic museums of the St. Mark's area. Although limited, this pilot scheme for managing the flows in conjunction with museum marketing will set an important precedent which, if successful, can easily be extended to other levels of flow regulation, such as transport, terminals, the hotel and restaurant sector. The importance of this scheme should not be underestimated. It represents a first attempt at managing the tourist flows by intervening in the supply by means of a policy of city marketing. Furthermore, this project is organised through a public-private partnership (PPP) which is a management instrument held to be particularly efficient and effective in harmoniously integrating the interests of the local actors involved (Van den Berg and Klaassen, 1988). Of course, the above mentioned ALATA project offers a wide range of opportunities to integrate those solutions in a whole tourism management support system of high profile.

A fundamental strategy for broadening the supply can be put into action in the following ways:

- giving value to the non-central attractions of the historical centre by constructing thematic routes and moving part of the museum supply out of the centre to flank the attractions which are already present there, with the aim of decongesting the central routes;
- giving value to the tourist supply of the surrounding region, such as the northern lagoon, Chioggia, the Brenta coast and the walled towns of the Miranese;
- giving value to and adapting the supply to non-urban segments of the tourist market such as that of seaside tourism, cruise or recreational tourism, naturalistic or green tourism;
- improving the quality of the existing hotel supply; as has been seen, despite the international reputation it enjoys as a tourist destination, today Venice still has very low accommodation standards with prices which are much higher than the national average. A first step towards engaging this strategy could be represented by the commitment to define and respect medium to high quality standards (by going beyond the designation of stars), but the opportunity to increase the number of low-budget hotel beds should not be overlooked;
- planning and spreading events over the span of the year. This should not be separated from their standardisation through the provision of automatic mechanisms for managing flows, access and information;
- the reorganisation and valorisation of the cultural supply, by means of the management and integrated promotion of the architectural and museum patrimony and the application of instruments for managing and spreading information on the logistics and the works exhibited which might allow a high increase in the commercialised product's value added.

2.2. Cultural tourism and the city's museum system[6]

If limited to the mere list of its cultural and exhibition structures in existence, Venice is a fully formed city of art; nevertheless, a critical examination of the tourist consumption of this heritage puts this point into doubt. Indeed, despite the high total capacity and the extensive artistic and museum heritage, a certain incapacity is perceived to attract the visitors to such cultural institutions and to its museums in particular. These show a low cultural and economic return which produces, first, the short average stay of the tourist who remains overnight in Venice, (2.3 nights against, for example, 3 nights for Florence), as well as, naturally, the very high number of day-trippers out of the total of visitors. Limiting our analysis to just the museum system of permanent exhibitions and to the temporary exhibitions which use these spaces, it is seen, firstly, how the bodies and levels of government on which they depend are extremely varied: these range, for example, from the State, in the case of the Accademia galleries, to the State owner but with city management in the case of the Doge's Palace (and not of other museums in the St. Mark's area), to the municipality in the case of Ca' Pesaro, to the clergy, to bodies of a mixed nature as in the case of the Biennial, to private foundations as for the Guggenheim Museum, as far as true and proper private industrial giants as in the case of Palazzo Grassi. The first consequence of this is the high diversification in promotion and selling strategies, the impossibility of pursuing uniform financial and cultural policies, and a badly defined role for the individual museums, which are often used both for permanent exhibitions and for temporary ones without a clear definition of their specific artistic propensity (Mossetto, 1993). In sum, a poor capacity for creating a comprehensive and autonomous cultural system which the policy for tourism development in Venice can be based on. This is translated into enormous potential problems of congestion, into a low profitability for the whole system and, more generally, into its low output in terms of cultural production, with serious repercussions namely on management costs and strategic significance.

The various analyses about the use of the cultural institutions make it quite clear that a link exists between visits to such institutions and the length of the trip. The question is quite simple: the Venetian cultural supply is so vast that it can satisfy the demands of a public with quite different preferences, if this public were adequately informed about *access*, had the possibility to book visits, could improve the visit's information content by accessing remote sources of further details, and could combine their own visit with opportunities for amusement and entertainment, etc. All this will be possible once the new information technology is able to back up the Venetian cultural supply in the form of a local network system, accessible on an international and world-wide scale. The possibility of using computers to raise the qualitative content and, consequently, the value of the cultural heritage is the object of a vast range of studies. Venice's opportunity is provided by the State telephone company which has chosen the city as an experimental cabled city.

The fundamental element for this operation will be the improved diffusion of such avant-garde technology to the tourist sector as a whole. Overcoming the barriers to entry would make it possible to draw on the positional rent on which such a large part of the Venetian tourist sub-economy lives, by increasing enormously the quality of the tourist experience had by visitors to Venice and by giving a new breath and a degree of authenticity to Venetian cultural production which is all too often reduced to trite clichés and to the *ad libitum* revisiting of a past which is increasingly distant from the real aspirations of the city and of its inhabitants.

This 'technological' scenario, it is clear, is neither an easy way out from nor a definitive cure for the consumption model which concerns the city at the moment. If, on one hand, it will facilitate the setting up of a production vocation which is, to a certain degree, independent of the tourist sector, thereby making it possible to confront some serious problems connected to the imposition of external costs on the city, it is not possible to imagine all the interest which Venice holds, only in terms of its cultural use. On the other hand, as Sassen (1994) maintains, 'telecommunications can support the change only if the cities have something to offer. The competitive advantage is the prime condition required, telecommunications are a secondary condition.'

2.3. Other segments of the tourist supply: congressional tourism, cruising, youth and elderly tourism, natural tourism

These emerging segments in the global tourist trends must be able to find adequate accommodation in Venice, both in terms of structures and the different quality–price mixes, and in terms of events and opportunities which could make it a priority destination. To turn Venice into an open city is a challenge which must be met, by equipping it with the necessary infrastructure (in the historical centre adequate structures for holding conferences still do not exist, a very important segment also from the perspective of the de-seasonalisation of the tourist flow), and by taking action for the training of human capital at various levels.

As to the necessity to diversify the local tourist supply, a reflection ought to be made on the opportunity to attract to Venice or to the coastal area those young people who currently prefer the seaside tourism of the Jesolo coast. Some attractions for this purpose can be the following: easily accessible beaches of a high quality (for example, provided with water-world type attractions); cinemas and audiovisuals of an IMAX kind or with virtual reality; structures equipped for ice-skating, swimming, diving, aquatic sports; music events in venues suitable for concerts at high voltage. In addition, there might be the possibility to meet the demand for isolation and for direct contact with nature of the green or natural segment, linked to special itineraries of eco-environmental interest such as the protected area of Valle Averto or, more generally, the lagoon area which contains various points of interest in terms of nature, landscape and folklore; theme parks; tourist villages linked to particular sports such as horse-riding, golf, sailing; theme hostels such as hotels for cycling tourists, archaeologists or naturalists; multi-use residences connected to the bed and breakfast service; school and work camps, and so on (Scaramuzzi, 1996).

In general, the objectives of a territorial planning policy which focused on the management of tourism would have to define a physical carrying capacity for goods and sites and a functional carrying capacity to be managed with resources and alternative utilisation methods.[7]

2.4. The policy for transport and access to the city

Here we want to focus on various key subjects which affect tourism, such as developing the airport, mobility between Venice and the mainland, diversification of access to Venice with the setting up of alternative terminals to the Piazzale Roma-Station one (to be restricted to urban traffic, the same as in the intentions of the new PRG for the historical centre.) This subject, in particular, excites great interest amongst the Venetian economic operators because it entails a redistribution of income advantages linked to the itineraries which connect these terminals with the more central districts and because it offers the planners the opportunity to revitalise the marginal areas of the historical centre. To this end, the plan drawn up by the municipality provides for the setting up (and the provision of adequate parking and information services) of at least three large tourist terminals to be connected to the landing-stages of the Arsenal, the Biennale area and the Zattere. Moreover, strategically locating some large museums precisely in these areas of lesser attraction (e.g. the museum centre of contemporary art and the one planned for the Salute area, as, also, the Arsenal's use for tourist and pleasure purposes, the natural-ethnographic itineraries of the northern lagoon) would encourage a positive effect of cultural production on the territory in terms of the economic dynamics already seen in the previous point.

As well as the already operating terminals of Tronchetto and Punta Sabbioni, there are two more terminals for access to the city used by visitors to Venice which are part of a project of rationalisation and adjustment of access. The first is that of Tessera, presently connected with Venice by means of road taxis, private and public bus services and water transport which includes a motorboat service which is the result of a partnership between the transport company (recently privatised) and a private taxi company.

The Fusina terminal, today of rather marginal interest, could regain its significance as a point of arrival for the flows coming from the South and the West thanks to the new town-planning measures which provide for moving the ring-road where the motorway exit for Milan is. Furthermore, the area's reorganisation is foreseen with the provision of a very large parking area and an area of service structures for tourists (restaurants, banks, shopping malls, information offices). The area, presently connected to Venice by a private shipping line, would have a service connection which would not function just as a ferry between Fusina and the landing-stage of the Zattere, but which is meant to form an independent and alternative route for the southern lagoon and exploit the island of the Giudecca's favourable position and expected reorganisation. This is expected to open the islands of the southern lagoon and the Lido coast to a tourist flow which is independent of the large itineraries of the historical centre.[8]

As far as the airport is concerned, an expansion of Marco Polo appears unavoidable in terms of the frequency of international flights which could take advantage (particularly where the charter sector is concerned) of an integrated management with the airports of Treviso, Verona and Trieste. The important subject of the environmental compatibility of water connections on the lagoon remains to be seen. These connections would become very busy; hence the potential need to excavate new shipping channels, with possible damage to the ecosystem of some of the richest areas, in biological terms, of the Venice Lagoon.

A more general consideration on the problem of accessibility to Venice remains. If, on one hand, it would, perhaps, be desirable to restrict as much as possible the opportunity to gain access to the historical centre, thus carrying out a selection at source of the flow of visitors (something which will only be partly possible once it is possible to regulate vehicle access by means of regulating access to the terminals with a booking and tariff system), the mobility of the citizens and, therefore, the need to diversify the access points according to the nature of the flows, remains a dramatic issue which risks threatening Venice's historical centre's position within the metropolitan economic circuits. From this point of view, the model of poor accessibility as defence against the tourist invasion can be ascribed to the 'vicious' scenario of subordinating the city's economy to the tourist economy. On the other hand, a model which prioritised mobility as a stimulus for reaching Venice from distant locations as well is fully inserted into a framework of the synergetic development of cultural tourism and of immaterial production activities with high value added.

2.5. Fiscal measures for the redistribution of benefits

In order to reduce the diseconomies created by tourist exploitation, the actions of urban management which are intended to be implemented would have to be accompanied by suitable fiscal measures for the redistribution of costs and benefits caused by the pressure of tourism. Out of all of the proposed measures, this is the one which involves a greater number of actors (the different levels of government) and which could face the largest political and administrative obstacles. Moreover, its effectiveness can only be assessed *ex-post* such is the uncertainty about the consumption choices relating to the use of tourist products.

There are two intervention philosophies in this sector: the direct or indirect taxing of the visitor; and the taxing of subjects who are not visitors but who benefit from the tourist consumption of the Venice destination in a 'derived' way. The first case chooses to make the visitor pay a price for the consumption of the city's services which cannot be taxed either through ordinary taxation, if not through indirect taxation, nor through tariff payments since it proves difficult and burdensome to differentiate between the different consumers. Indeed, whilst in the case of public transport it is simple to differentiate and to make the tourist pay a higher price, it would not be as simple to do so for the use of roads, of the images, of communal beaches, of waste collection services, etc. The most obvious opportunity for collecting this tax seems to be that of the payment of the price of the overnight stay: all those who stay in Venice could be asked to pay a higher price which would be assigned to covering these costs (the so-called 'tourist tax').

The problem with this kind of indirect taxation is that it hits the tourists who stay overnight exclusively, discriminating in favour of day-trippers and thus obtaining the opposite result to the one desired. Furthermore, this kind of taxation is unpopular with hotel owners who fear a reduction in their profits.[9] A tax of this kind was in force until a few years ago, and was subsequently abolished. A drastic solution would be to make all visitors pay an entrance fee at the access terminals but this would have an effect which is negative in many respects, both because of the difficulty of running a checking system at the access points which could distinguish between those who come to Venice for tourism and those who come for work or study, and because of the difficulty of avoiding the constitutional norm which guarantees all citizens freedom of movement within the national territory. A further difficulty would concern the city's overall image which finds itself faced with increasingly aggressive international competition.

An alternative and more promising system is that of making a fiscal redistribution between the places which are the destinations of the day-tripper flows and the places which generate them, with the aim of resolving the spatial imbalance between costs and benefits generated by tourism. There are two possibilities for carrying out such a redistribution:

- On the basis of *ex-ante* theories on the capacity for generating day-tripping towards a given destination. With this model, it is presumed that every municipality (or province), whatever the effective spillover that it imposes to the city which is the destination of the daily flows, generates a number of day-trippers towards that city which is inversely proportional to the distance and directly proportional to the overnight-stay tourists who are there. An exemption from these transfers could be envisaged for all the municipalities (or provinces) which, on the basis of these two parameters, do not reach a pre-established minimum revenue. This method has the advantage of not requiring ad hoc surveys, since the data on overnight stays collected by the tourist promotion companies suffices. However, this would risk being unmanageable even if a kind of tourist chamber of compensation were organised, similar to that of the banking system, which determined centrally the net transfers as the balance between active and passive values.
- In order to get around this obstacle, it could be better to focus on the large destinations of cultural tourism, those most hit by day-tripping and those most easily identifiable, and to impose the transfer system on the basis of *ex-post* data on the place of origin of the day-trippers who go there, which should not be impossible to collect with periodic surveys.

Whichever system is chosen, it must be accompanied by further local redistribution measures which are more effective than the present one, also by means of the fight against illegal activity and tax evasion which are extremely high in the tourist sector, especially where small enterprises and individual firms are concerned. Certainly, the prospect of a serious administrative and fiscal decentralisation will help Venice, just as any other tourist city, to design the fiscal system which is most suitable for its own specificity and its own problems.

2.6. A Venetian authority for tourism

Tourism is a transversal activity which involves an extremely vast range of authorities, local and otherwise, and many enterprises belonging to different sectors. The consequent fragmentary nature of competence and of interests makes tourism a sector which is difficult to manage.

Van den Berg *et al.* (1996) have identified the organising capacity as the key to the quality and competitiveness of the total tourist product. The management of tourism today is much more than its simple promotion, and it involves co-ordination between many public and private actors. To complicate this picture, there are the spatial features of the phenomenon: demand is geographically dispersed whilst supply is concentrated. This means that many important interlocutors, for example, tour operators, remain outside the administrative competence of those who manage the tourist resources.

An answer which is often proposed for these types of management problems is the creation of an Authority for Tourism. This Authority, analogous to that which has been set up for the port, for example, ought to assemble appropriately public and private interests and propose a policy of integral management of tourism to the city, taking the interrelations between the different subjects involved into consideration.

3. Final considerations: the challenge of sustainable tourism

Sustainability is a concept which is extremely ambiguous and fuzzy, even though, recently, there has been a widespread effort to reach a general and shared definition for it. Such a definition is provided by the Brundtland Report, 1987 (published in the volume *Our Common Future*): 'sustainable development is that development which satisfies the needs of the present generation without compromising the capacity of future generations to satisfy their own needs'. A more specific definition was drawn up by the ICLEI (International Council for Local Environmental Initiatives): 'sustainable development is development that delivers basic environmental, social and economic services to all, without threatening the viability of the ecological and community systems on which these services depend'. There are two key concepts: the delivery of services as dependent on the systems; the community involved.

Evidently, placing such definitions in the economic, social and cultural context which is the subject of the study, they must be given appropriate meanings. The debate which accompanied the Fondazione ENI Enrico Mattei's *Progetto Venezia* emphasised many times the danger of taking a notion of sustainability which is uprooted from a historical and social context and, moreover, without scientific foundation for granted. For example, how must the reference to future generations be understood in Venice's case? Which future generations are being referred to? The children and grandchildren of today's Venetians, or of that great mass of potential Venetians, especially the university students, which in some way have to be persuaded to become Venetian citizens in order to slow down the depopulation trend in progress? Or even the citizens of the world, since the city of Venice represents a historical heritage of such importance that is perceived to belong to the whole humanity?

Does a development's cultural sustainability exist, or, rather, are a city's identity, image and social characteristics dependent variables in a dynamic development process which is based on purely economic elements? In other words, must Venetian development respect the constraint of its identity as a historical city tied to international movement, to artisan activities and to tourism, or can it develop outside this paradigm, attracting new production activities which are apparently incompatible with its historical image of boats and stone.

Sensitivity to these issues is changing, opinions differ, but it is a fact that the increasing drive of socio-economic pressure factors has recently provoked a gelling of these positions on the city's projects, finding their most obvious expression in discussions on the new town-planning scheme (the PRG for the historical centre). Today, long-term vague visions are no longer talked about: it is a shared idea that in the medium- to short-term the local community must make fundamental choices for its future, otherwise it risks remaining marginalised from every possibility for growth and being bound to an urban model which is increasingly exposed to the global processes of modern society.

The idea which is being asserted of a sustainable city is that of a historical centre which is fully integrated within a diversified metropolitan context, in which Venice plays a fundamental role by specialising as a representative place, cultural attraction and seat of government. Furthermore, at the same time, it acts as a hub of wide networks such as the regional network of Eastern European trade, the network of European cities of art, or the virtual telecommunications network, in relation to which, thanks to its strategic position in the first case, and thanks to the elimination of distance in the second, Venice would gain a new accessibility and a renewed strategic position. It is not up to the authors to decide whether this idea of the city is fully sustainable from every point of view. What is certain is that an initial and fundamental step towards sustainability in the economic sense would be taken if this path was definitively taken up, with the necessary steps from the institutional-administrative point of view as well.

Strictly tied to the issue of the social and cultural sustainability of Venetian development is the problem concerning the full representation of the interests of Venetians. Are Venetians today in a position to best decide about the future of their city? Or are they themselves not, perhaps, the stakeholders who are most bound to a conservative decision and, as such, an unsustainable one? By their own admission, the city's present governmental organs often have to take decisions which leave the current preferences of the local population out of consideration and which relate to a much broader potential demand, both in a temporal and a geographical sense. This demonstrates the inadequacy of local institutions (but also national ones) for managing medium-term processes and the need to study governance models which are innovative and widely acceptable.

The picture of problems and possible evolution concerning the Venetian economy and, in particular, its leading industry, or the tourist economy, must be assessed with the methodologies adopted by the municipal authorities as a result of the Agenda 21 process. This provides for a participative process for public decisions based on quantitative and qualitative classification instruments, i.e., indicators for sustainable development. After making the policies clear (on which, in any case, a certain degree of consensus exists: the proposal to introduce a Venice Card for regulating the tourist flows was supported by the great majority of the political parties engaged in the administrative elections of 1997), the monitoring stage becomes fundamental in signalling a break with the past and in conferring continuity and feedback on the process.

Obviously, the conflict taking place over the use of the city which is most clearly perceived by the public, by the media and by local politics, is that which concerns its tourist function. In this sphere, as is no other, it is difficult to make courageous decisions because to embark upon the path towards sustainable tourism requires dealing with consolidated interests immediately, 'revealing' preferences for the city's future to the actors who, until now, have remained somewhat excluded from the political debate, rethinking the city's policy with completely new creativity and entrepreneurial capacity. But this does not mean that sustainability must leave winners and losers in its wake.

Indeed, the sustainability challenge, in the authors' opinion, lies completely in the creation of a quality tourism which is suitable for all of the economic and social actors involved in the process of the city's economic development, including the subjects who, today, contribute most to the decline in quality and the short-sighted preference for quantity. It could prove quite arduous to venture as far as making projections or simulations on what the trend of the macroeconomic variables looked at in this research could be in this scenario which, today, appears futuristic in many respects. Nevertheless, we believe in the possibility of creating, henceforth, a monitoring campaign which could help the public decision-makers to take opportune action in the directions indicated in this chapter, thereby minimising the possibility for error in the strategic decisions made and conferring a democratic and unified nature to the whole process.

Notes

1. See Van der Borg and Russo (1997) for an exhaustive description of the main trends which have characterised Venetian tourism in recent decades.
2. See also Indovina (1988).
3. These values can be subjected to a new estimation taking both the changed propensities in the city's use (redistribution in favour of the amount of day-trippers) and the variations in the considered constraints into account. Furthermore, it would be very interesting to take environmental constraints into account explicitly.
4. A further result provided by this model is the identification of severe constraints on which a policy of broadening supply must be based in order to mitigate the pressure exerted by tourism on the city. For example, while the value for the constraint on water transport suggests that an

improvement in capacity, meaning a differentiation in tourist routes, can make it possible to increase revenue from tourism, increasing the waste collection capacity or the number of parking places would probably be of no use for this purpose.

5. These correspond, in a broad sense, to the categories of state, pressure and response indicators, which are usually used to describe the state and trend of a territorial system, even though an effort to create site-specific indicators has been made, in accordance with the methodology shown by the World Tourism Organisation (cf. *Consulting and Audit Canada*, 1995). For the methodology used in creating such indicators and the values monitored by using them, see FEEM 10–97.
6. Much of the data shown here is taken from S. Funari, I. Simionato, (1996).
7. Scaramuzzi (1996, p. 234) talks of 'minimum thresholds – or maintenance capacity – under which disuse or – worse –compromise of the good takes place, to the point when it is definitively lost'.
8. COSES estimates the hourly flow of passengers to the historical centre which comes from the 29 to 146 tourist coaches and from the 473 to 2400 cars a day, according to the season, with a number of journeys required for the full functioning of this terminal which, in rush hours, is equal to 1 motorboat every 10 minutes in summer time plus 3 extra journeys during the maximum period in September.
9. Actually, the demand for tourism services is decidedly inflexible and therefore the tax is almost completely transferred onto the final users (this hypothesis is confirmed by a study with ex-post data on a beach resort like Hawaii, with an estimated transfer of almost 100%: see Bonham *et al.*, 1992). This will apply even more for a destination with a hotel supply which has features such as Venice's.

Bibliography

Ashworth, G.J. and Voogd, H. (1990) *Selling the City*. New York: Belhaven Press.

Associazione Venezia 2000 (1993) *L'accessibilità a Venezia: le tre stazioni, S. Lucia, Marittima e Tronchetto*, Dossier/1, Venice.

Bonham, C. *et al.* (1992) The impact of the hotel-room tax: an interrupted time series approach. *National Tax Journal*, **45**, n. 4.

Bonifacio, M., Vassallo, M. and Zanus Fortes Gaia, L. (1995) *Indagine sulla condizione abitativa degli studenti fuori sede*: *anno accademico 1993/1994*. Venice: Ufficio Studi e Programmazione' of the University of Venice.

Butler, R.W. (1980) The concept of a tourist area cycle of evolution: implications for management of resources. *Canadian Geographer*, **24**, n. 1.

Carraro, C., Costa, P. and Manente, M. (1994) *Il modello STREP di previsione dei flussi turistici internazionali in Veneto: struttura teorica, parametrizzazaione e previsioni al 1995*. CISET Working Paper, n. 8.4/1994.

CISET (1996) *Applicazione delle Telecominicazioni e della Telematica per la Gestione Ottimale dei Flussi di Informazione nel Settore Turistico*: research commissioned by the Centro Studi Telecom Italia S. Salvador.

CISET/GRETA (1997) *1997: Performance Ancora Incerte per il Sistema Turistico Italiano: il Movimento Internazionale da e per l'Italia Previsto dai Modelli TRIP*. Report prepared for the conference organised by the Italian Touring Club for the BIT, Milan, 28 February 1997, edited by Carraro C., Costa P. and Manente M.

Cogo, V. (1997) Gli indicatori di sostenibilita':primi risultati di una ricerca per Venezia, Fondazione Eni Enrico Mattei, Working paper 10.97, Venice.

Comune di Venezia (1996) *Piano per lo sviluppo e adeguamento della rete di vendita, Relazione di sintesi degli studi preliminari al piano*, Venice.

Consulting and Audit Canada (1995) *What managers need to know: a Practical Guide to the Development and Use of Indicators of Sustainable Tourism*. Report prepared for the WTO.

COSES (1978) *Turismo a Venezia*, Venice.
COSES (Ed.) (1996) *Alfabeto veneziano 1996: economia e società nell'area metropolitana veneta*. Bologna: Il Mulino.
Costa, P., Manente, M. and Van der Borg, J. (1993) *Tourism in Traditional Tourism destinations*, CISET Working Paper 93.1, Venice.
Costa, P. and Van der Borg, J. (1988) La determinazione della capacità di carico turistica di Venezia. COSES *Informazioni*, vol. 32/33.
Di Monte, G. and Scaramuzzi, I. (Eds.) (1997) *Una Provincia Ospitale*. Bologna: Il Mulino.
Funari, S. and Simionato, I. (1996) I musei veneziani: indagine sulle strutture permanenti. In: Roncaccioli , A. (Ed.) *Azienda Museo*, Padova: CEDAM.
Glasson, J. *et al.* (1993) *Visitor Management*. Oxford: Oxford Brookes University.
Indovina, F. (1988) Turisti, pendolari, residenti. *COSES Informazion*, vol. 32/33.
IUCN (1992) *Guidelines for Protected Areas Management Categories*. Gland, Switzerland: International Union for the Conservation of Nature and Natural Resources.
Kotler, P., Haider, D.H. and Rein, I. (1993) *Marketing Places*. New York: The Free Press.
Lando, F. (1986) Functional areas of the town: the example of Venice, Italy. *Urban Economics*, 13.
Manente, M. and Minghetti, V. (1993) *I modelli di previsione del movimento turistico da e per l'Italia: la banca dati*. CISET paper, n. 5.
Manente, M. and Rizzi, D. (1993) *I flussi turistici a Venezia 1989–1992*. Working Paper. Venice: Department of Economic Science.
Manente, M. and Scaramuzzi, I. (1995) The Case of Venice. CISET paper presented at the conference *City tourism in central Europe*, Ljubljiana, 12–14 October 1995.
Mossetto, G. (1992) *L'economia delle città d'arte*. Milan: Etas Libri.
Mossetto, G. (1993) An alternative route in Venice. In: *Alternative Routes in the Cities of Art*. Technical Report no. 23. UNESCO/ROSTE..
Onishi, T. (1994) A capacity approach for sustainable urban development: an empirical study. *Regional studies*, 28, n.1.
Pearce, D. (1989) *Tourism Development*. New York: Longman Scientific and Technical.
Pedenzini, C. and Scaramuzzi, I. (Eds.) (1997) *Commercio e città: un laboratorio per il piano commerciale di Venezia*. Bologna: Il Mulino.
Preiti, A. (Ed.) (1994) *Turismo Domani: la Sfida degli Standard*. edited by Venice: Associazione Venezia 2000.
Prud'homme, R. (1986) *Le Tourisme et le Developpement de Venise*. Venice: Mimeo.
Rispoli, M. and Van der Borg, J. (1988) Più vicini, più cari. *COSES Informazioni*, 32/33.
Rullani, E. and Micelli, S. (1998) *Immaterial production in Venice: towards a post-fordist economy*, in this volume.
Russo, A.P. (1998) *Organising Sustainable Tourism Development in Heritage Cities*. Working Paper EURICUR, 98/02, Rotterdam.
Sassen, A. (1994) *Cities in a World Economy*. Thousand Oaks, USA: Pine Forge Press.
Scaramuzzi, I. (1996) Piano territoriale provinciale: una modesta proposta per il turismo. In: COSES (Ed.) *Alfabeto Veneziano: Economia e Societa nell'area Metrolititana Veneta*. Bologna: Il Mulino, pp. 239–240.
Tinazzi, S. and Van der Borg, J. (1991) Il sistema turistico veneto: un'analisi territoriale. *La Rivista Veneta*, 38.
UNESCO/ROSTE (1993) *Proceedings of the International Seminar: 'Alternative Tourism Routes in the Cities of Art'.* 24–25 June, Venice, Italy. 1993 *Tourism and cities of art*, Technical Report n. 20 (ed. van der Borg, J.).
Van den Berg, L. and Klaassen, L.H. (1988) *The role of public-private partnerships in the process of urban revitalisation*. Rotterdam: Netherlands Economic Institute.
Van den Berg, L., Van der Borg, J. and Van der Meer, J. (1995) *Urban Tourism*. Aldershot: Avebury.
Van der Borg, J. (1991) *Tourism and Urban Development*. Amsterdam: Thesis Publishers.
Van der Borg, J. and Russo, A.P. (1997) *Una Sistema di Indicatori per lo Svilippo Turistico Sostenibile a Venezia*. Working paper n.06.97. Venice: Fondazione Eni Enrico Mattei.

Van der Borg, J. and Russo, A.P. (1997) *Lo Svilippo Turistico a Venezia: Analisi Territoriale e Scenari di Sostenibilita*. Working paper n.07.97. Venice: Fondazione Eni Enrico Mattei.
Var (1992) *Il Turismo a Venezia Verso il 2000*. Venice: University of Venice.
World Tourism Organisation. *Tourism Highlights*. Madrid, various years.

CHAPTER 7

Immaterial Production in Venice: Towards a Post-Fordist Economy

Enzo Rullani and Stefano Micelli

1. Four scenarios of sustainable development for Venice

Planning in Venice must consider the question of sustainability. The evolutionary courses which are produced by policies define the scope for the possibilities or all the alternatives which can, in theory, be pursued. In order to be agreed upon and credible in terms of society's debate, the alternatives must shape development courses which are really sustainable. That is, development courses which can, on the one hand, sustain themselves by rebuilding, over time, the reasons as well as the resources consumed in development and, on the other, courses which can regenerate the city's identity in a continuous way, by re-elaborating it in gradually different forms which are adapted to the circumstances which emerge in the evolution of the environment and of the urban system.

Only if these two conditions are satisfied can we say that the course of development for a city, in our case Venice,[1] is sustainable. In other words, by following this route, the city can grow, can change, can live and, above all, can remain 'itself'. What would not be sustainable is development which instead, by not satisfying the first requirement, were to impoverish the city by exhausting the historical or natural resources on which the development is based or by causing the institutional, cultural, and social basis for the growth to break down. In the same way, any development which were to radically change the city's identity, thereby making it impossible for the new city to reflect the old one, would not be sustainable either.

The concept of sustainability takes on connotations which are much more complex than those which apply to material sustainability alone (replacement of consumed material resources). Indeed, a city is, above all, a system: a sum of social and institutional coherence. Applied to the city as a system, the

I. Musu (ed.), Sustainable Venice, 195–226.

concept of sustainability must relate not only to the reproduction of the material resources consumed, but also to the reproduction of the conditions for the system's social, institutional and characteristic coherence. Thus, a term of reference is introduced, sustainability, which is fairly ambiguous but which is necessary in order to avoid trivialising the city by reducing it to a basic aggregation of resources. Therefore, a project for Venice must also contain the following elements: the engendering of a shared vision of the city as a system which is not only a material system but also a social one, as well as the conditions which could permit the reproduction or, rather, a new elaboration of its memory in time.

To maintain the capacity to recognise over time its own identity as an urban system is an important requisite for sustainability. It is a requirement which must also hold good in moments of changes, however important, in the physical persons which make up the city, in the activities carried out there, and in the structures and institutions which organise city life. A city which lost its sense of history would destroy the possibility of being shared by its inhabitants. Without a historical identity these inhabitants would become, in fact, a multitude of separate individuals, involved in competing on different markets, but without common values. The city which, in order to enrich itself economically, loses its identity along the way, in reality dies: it disintegrates into its constituent elements and stops existing as a coherent and integrated system of values, actions and dependencies.

There are various courses for sustainable development in Venice's future. The base-course is the one established by inertial evolution, without any deliberate intervention. Scope for possible courses has been formed by the simulation, for now only theoretical, of action (the actors' strategies and/or public policies) which is able to divert evolution from its 'natural' course. For each of the main options for intervention on the base-scenario, three alternative scenarios are obtained. Each of these constitutes a kind of course which is, in abstract terms, coherent, sustainable, and has advantages and disadvantages which are not equally distributed, to the benefit or detriment of the different interests at stake. The analysis which will be carried out for each of these scenarios does not aim to advance one of them as the best, but points out the differences between them and defines the criteria for their internal coherence.

Obviously, real evolution will not correspond to any of these courses in their pure form as they are presented here. In political and operational practice, there will be compromises, mediations, and integration. It is precisely for this reason that the shared recognition of the differences which characterise the scenarios considered is necessary.

The scenarios selected in order to define the scope of the possible alternatives are the following:

- the inertial growth scenario, spontaneous, without intervention;
- the radical environmentalism scenario which could be brought about with a policy which assumed the environmental constraint radically and which adapted the whole evolution of the city to such constraints;

- the metropolitan integration scenario which assigns Venice the role of a functionally specialised district of a metropolitan city which includes Mestre, Padua and Treviso (PaTreVe)[2] and which extends to the whole of the Northeast for certain functions (the port, the airport, the cultural institutions);
- the accelerated transition to post-Fordism scenario, which sees Venice as the specialised link in a network which is not only, or principally, metropolitan but which rapidly becomes the global network as a result of the widespread use of new digital communication technology.

In all four cases, immaterial production, which already plays an important role in the city's economy, would act as a driving force in the urban system's evolution. However, there would be different types of immaterial production.

However, let us examine the four scenarios.

1.1. Inert growth

The inert scenario is the one which, also considering the difficulties encountered by policies for intervention, is most likely to take place. This continues along the course which already exists at the moment, without any great innovation. It is divided into the following parts:

- an increase in Venice's specialisation in tourism, since tourism is the only activity equipped with extended networks (to the global limit) and the only one capable of generating value even in the absence of policies for innovation;
- Venice's growing separation from the mainland and Mestre's integration into the metropolitan area. In this scenario, the functions presently carried out by Venice will tend to be transferred to Mestre or, in some cases, to Padua or Treviso (e.g., universities, banks, hypermarkets, health services, public administration);
- Porto Marghera's internal evolution, no longer tied to Venice in terms of functions. In this scenario, the progressive dismantling of Porto Marghera's heavy industry can be foreseen: without a politically significant undertaking for the recovery and organised re-use of the industrial areas, the most probable result in the spontaneous scenario is the partial repopulating of the area with small industrial activities which could benefit from using the existing infrastructure.

Is this a scenario which is sustainable in the long-term?

Undoubtedly, there are problematic elements which lead to a consideration of the possibility of a serious decline in the production of value in Venice and of a perversion of the city's true nature in terms of its identity which is difficult to restore. Indeed, the advantage which PaTreVe would have over the historical city in terms of metropolitan attraction could only be offset by giving a free rein to the historical centre's exploitation by tourism.

1.2. The radical environmentalism choice

Venice has threatening environmental problems. These are partly connected to age-old imbalances (long-term subsidence, increase in the average sea-level) which could affect the relationship between the submerged part of the lagoon and the parts above sea-level. This is a relationship which already provokes enormous uncertainties in terms of the policy for curbing high water (the Mose project). In addition, the problems related to the erosion of the lagoon bed and to the transformation of the biosphere linked to the increased salinity of the waters must be considered: the *Canale dei Petroli* (oil-tanker channel) seems to be the main cause. To complicate the picture, the equilibrium of the lagoon has been altered by the location of heavy industry in the areas of Porto Marghera.

In the majority of cases, these are problems which are difficult, if not exceedingly difficult, to solve. Moreover, for this reason, it is unlikely that this scenario could be confirmed as a radical choice which is destined to endure against other demands and other interests at stake. Nevertheless, the radical nature of the environmental constraints are objectively compulsory for Venice and justifies placing this type of scenario in the definition of the scope for possible evolutionary courses. Indeed, the possibility must not be ruled out that a significant accident could take place, or that worrying discoveries could be made on the environmental degradation and on health risks, or that accidental events could occur, even outside the lagoon, all of which could have an impact on the sensitivity of public opinion, and cause absolute priority to be given to the environmental constraint, adapting the whole Venetian economy to it.

In this case, the foreseeable scenario is dictated by a massive shift in the sources which generate value. The development of new activities in the field of the environment (recycling, recovery, fishing, environmental tourism, etc.) would compensate, but only partly, for the loss in income and in employment which would occur in heavy industry and for the greater costs sustained by urban logistics.

The result of a policy aimed at objectives of this nature is easy to predict: there would be a progression towards the accelerated shut-down of the works which pollute the most and which require the largest tonnage (refining, petrochemicals) as well as the fast decline of the commercial port. The consequences of all this seem extremely important. First of all, there would be consequences on the whole logistic system of the Northeast: with the refinery closed and also, potentially, the petrochemical industry, with the commercial port closed to the transit of large tonnage ships, with a limit to the movement in the lagoon of ships containing pollutants or oil products, the petrol and diesel supplies for the central-east Padanian plain, which are presently transported from Venice, would have to be diverted elsewhere (the ports of Ravenna and Trieste seem to be the main candidates for replacing Venice in this function.) The reduction in the level of the commercial port's and the industrial port's activity entails the loss of the main factor for forming the complementary relationship between Venice and the Northeast. The logistics of the businesses of the Northeast would have to go elsewhere, bypassing Venice as, after all, has happened so far.

To make up for this, it would be possible, with appropriate investment, to give value to tourism of an environmental nature, oriented towards tourism's utilisation of the lagoon environment. Environmental recovery and recycling could create earnings and jobs which would, partly, compensate for the losses sustained. The lagoon could be exploited as a productive resource even if, to respect the requirements of environmental protection, the economic exploitation of the waters must be carried out within precise limits which must not be overstepped.

Is this a scenario which is economically sustainable? Probably yes, in the long-term, even though it would involve making heavy sacrifices in the short-term: a consistent diet would be forced onto a city which has already remained behind, in comparison to other areas, in generating wealth and creating jobs. Both Mestre and Venice would suffer the loss of earnings and employment deriving from the increasing restriction of Porto Marghera's activity. In the Northeast of 'miracles', Venice as a whole would end up as being an area in crisis or, at least, an area of reduction in income and employment. It must be understood that this is by no means dramatic: Mestre could compensate for the loss with a greater integration with the surrounding municipalities of a higher industrial density, by selling them services as one of the central areas of PaTreVe. Venice could, instead, look for compensation in an intensification of tourism.

1.3. Metropolitan integration

This is the scenario most directly concerned with constructing a city which systematises the present urban segments (the historical centre, Mestre, Porto Marghera) by maximising their potential for specialisation and integration.

Those who identify metropolitan integration as the point of departure for a social and economic revitalisation of the historical centre view the issue of physical accessibility to be the distinction between the present and the past and between scenarios for preservation and those for renewal. In other words, all of the projects for developing the city take the end (or at least the slackening) of the logistic constraint which currently separates Venice from the mainland for granted. Without a removal of this structural boundary any plan to locate in Venice will be subjected to costs and disadvantages which no economic or institutional operator can allow itself to sustain. Whilst the tourist who comes to Venice once in a lifetime can find the present division of local transport (as well as the various changes required during the journey for reaching the historical centre) and the difficulty in reaching certain parts of the city entertaining, the resident considers all of this as unproductive cost. Only by going to the root of the problem is it possible to create the basis for a greater differentiation within the Venetian economic system, by envisaging development in sectors alternative to the tourist sector.

The keystone of this new integration is the project for light railway, an infrastructure for mobility which should grant access to Venice's various districts with modes and timescales which are consistent with constraints in the rest of the neighbouring area.

The prospect of metropolitan integration tends to heighten the specific nature of the various urban segments. Porto Marghera can preserve its heavy feature, linked to its basic industry (industrial port) and to the sea transport of poor goods (commercial port). In this scenario, Mestre takes on the role of the hinge-city placed, as it is, in the centre of PaTreVe. This is an important role, even though, for now, it is more a role of making a connection between centres which remain independent than one of attracting the different nuclei of PaTreVe towards an central point of reference. Even the historical centre is destined to strengthen its specialist vocation of a city suitable for activities of a light, immaterial nature which require limited commitment on the material side: tourism (but not exclusively), culture, art, communication and headquarters.

All of these activities could be given a powerful boost as a result of improved contacts with a market which is expanded to the metropolitan area. In some of these fields, Venice could supply demands located in Mestre, in Padua or in Treviso. At the same time, Venetian residents would have the possibility to use the services available to the whole metropolitan area without great difficulty. Furthermore, this could considerably improve the likelihood of residence in the historical centre, once the quality services also become accessible from the historical centre.

1.4. Accelerated transition to post-Fordism

In the metropolitan scenario outlined above, global networks and the production modules' territorial mobility play a relatively minor role. More than anything else, metropolitan integration assembles all the things which exist, making the links and the connections denser and changing the physical infrastructure which supports mobility and transport. The hypothesis which the construction of this scenario is based upon is that the problem of developing the historical centre is strictly related to the physical accessibility of island Venice. Furthermore, that its unification with the metropolitan area (or, rather, a removal of the structural restraints which act as obstacles in the present system of transporting people and goods in this area) could result in substantial benefits on account of a new specialisation trend. Therefore, the creation of an integrated system which might see Venice as a constituent part of the metropolitan area, and not simply one of administrative reference, makes it possible to imagine forms of specialisation within a market from which the historical centre remains essentially excluded today.

However, the scenario can change, with some surprises also in store, if the prospect of a transition to post-Fordism is henceforth considered to be a present-day prospect. In this case as well, the hypothesis of a scenario for developing the historical centre starts with a plan for transforming conditions for accessibility to the historical centre of Venice. In this case, the problem is not posited in physical terms (mobility and transport) but, above all, in information terms (information technology networks). In operational terms, this means:

- Basing the division of labour not only on metropolitan networks but also, at the same time, on global networks which transfer codified knowledge and make a division of labour possible on a world-wide scale;
- Raising the quality of institutions and businesses which operate on line with the aid of co-operative work instruments supported by computers in such a way as to encourage forms of work carried out within networks which can guarantee the co-operation of specialists on a national and international scale;
- Developing a programme of computer literacy for the population which could grant access to sophisticated forms of participation and to the use of the new means of communication to groups who are presently excluded from services involving information technology (e.g. the elderly, unemployed, housewives).

The progression to a post-Fordist economy is characterised above all by a new culture of demand. It is not necessarily a matter, as several international observers have recently pointed out, of bringing programmers from all over the world to Venice for the purpose of making the city a capital of software or of electronic component production. Instead, it is necessary to reconsider the entirety of the activities currently carried out in the historical centre in different terms, by evaluating, case by case and sector by sector, the impact of the new information technology on the transformation in the chain of value in the different economic functions. Each immaterial activity present today in Venice, from training to trade, from the supply of museum services to publishing, is open to being digitalised and made virtual. What it is necessary to do is to make an accurate evaluation of each activity's competitive potential in relation to a scenario which changes the geographical dimension dramatically and which tends to be characterised immediately as global.

In a context of accelerated (and politically forced) development of information networks in the Venice area, and particularly in the historical centre, the metropolitan integration of the preceding scenario would still be valid, even if its power over the functions allotted to the different districts of the metropolitan space would be considerably mitigated. Indeed, each area would have a better possibility of freeing itself from the division of labour made on a metropolitan scale, to find other opportunities of exchange and of integration. The districts would no longer be defined by their one or few vocations but would come to have far more differentiated activities which belong to different networks and which overlap each other on the same segment of the territory.

In this kind of scenario, the locating constraint counts less than subjective capacities. The possibility of entering into the virtuous circle of specialisation which nourishes itself by means of the networks depends on personal willingness, which not all have, for making an initial investment in learning (computer literacy), in infrastructure (information technology networks), in organisation (relationship networks and networks of trust). These are the patterns with which these investment choices mature in determining positions in the transition to post-Fordism.

In this sense, a highly unpredictable situation opens up in which the personal strategies of the businesses and/or the cities can overlap. In other words, it becomes difficult to make forecasts: more difficult than in the preceding scenarios.

Nevertheless, the greater unpredictability also corresponds to the increased freedom which the post-Fordist scenario grants a city which, until now, has been strongly restricted, for better or worse, by its historical and natural conditioning. So, without denying its nature of a city on water, Venice could fully participate in global networks which can overcome the city's physical isolation and reopen opportunities lost during the years of Fordism. The sustainability of this kind of scenario is linked, above all, to the extent to which the personal strategies and political intervention can really accelerate the formation of a multimedia culture in the historical city, thereby compensating for the slowness which can still be seen today in experimentation and innovation in this field.

If this acceleration takes place, thereby bringing Venice back into the large global networks, there is no doubt about the final result: Venice could return to being a significant player in global networks. However, everything depends on the speed of this transition: if it is slower than that shown by other competing cities (also by Mestre, Padua, Treviso), then there is the risk that the Venice area would remain relatively impervious to the new knowledge networks and that the activities which are most affected by the multimedia revolution might be moved elsewhere. Thus, Venice could be left in a state of marginality which could no longer be remedied.

2. The immaterial economy in Venice

2.1. The make-up of the immaterial economy in the municipality of Venice[3]

All of the scenarios considered refer, both explicitly and implicitly, to the immaterial economy in Venice. In this section, we intend to analyse immaterial production in the municipality of Venice, beginning with an examination of the data obtained from the recent censuses taken by Istat.[4] We shall begin by making a general survey, with the aim of providing some elements of basic knowledge.

In order to study the Venetian territory's immaterial sector, it is beneficial to begin by assessing the overall composition of the sector.[5] An initial look at the data collected immediately reveals that the immaterial economy constitutes a significant part of the total economy of the municipality of Venice: those employed in the sector number 62,521, comprising 45% of the total employment in the municipality (139,060). The local units number 11,486 out of a total of 18,693 (61%) (Tables 1 and 2).

The amount of employment absorbed by the other service activities (tourism, transport, non-retail trade) is much smaller, almost 37,792 employees, or 27% of the total, similar to employment in industry (38,747). The local units account for more than the 3000 units of the industrial sector and are around 4000 where tourism, transport and wholesale trade are concerned.

The economic activities which can be classified in the immaterial category are different. The most numerous group is formed by the retail trade division which has 15,026 employees and 5602 local units. Second in order of size, with more than 10,000 employees and 2200 units, is the compound section K of the ATECO 91 classification (summarised in the table under services for businesses), which includes real estate and rental activities, activities connected to information technology, professional and entrepreneurial activities (legal and notary firms, business consultancies, etc.). Each of the services for the community (public administration, health, education) have around 8–9000 employees. The other services have more units (1544) but fewer employees (4952), while the banks and insurance companies total 377 units and 4398 employees. The least important sector in quantitative terms is publishing which has only slightly over 900 employees and 150 units.

This was the situation in 1991. In order to give meaning to these figures, it is necessary to make a comparison with the corresponding figures for 1981, in order to evaluate the trend between the censuses, and with the relative specialisations of neighbouring cities such as Padua and Treviso. In this way, it is possible to discern, on the one hand, the sectorial trends in the 1980s in order to assess which sectors have undergone forms of development in this 10-year period and which have experienced forms of rationalisation and reduction in employment. It is also possible to compare the trends of growth and reduction in employment in relation to the trends seen in the two cities which, together with Venice, make up the metropolitan area called PaTreVe. This area, an integral part of the system of the Northeast, which today has been promoted as being a model for economic development at a national level, provides the terms of comparison on which to evaluate the real trend of the economy of island Venice and of its immaterial component.

An appraisal of the real possibilities for developing an innovative tertiary sector in Venice which could project the city into the national and international economy by attracting new employment opportunities and new residence possibilities requires, first of all, a comparison with the vocations and capacities actually shown by the economies of the other municipalities which make up the metropolitan area. It is a question of assessing what the real trends in progress are and what the boundaries within which potential intervention can take place are in relation to evolutionary tendencies already consolidated.

Table 1 Local units (L.U.) and employees in the municipalities of Venice, Padua and Treviso in 1981 and 1991

	1981						*1991*					
	Venice		*Padua*		*Treviso*		*Venice*		*Padua*		*Treviso*	
	L.U.	*Emp.*	*L.U.*	*Emp.*	*L.U.*	*Emp.*	*L.U.*	*Emp.*	*L.U.*	*Emp.*	*L.U.*	*Emp.*
Immaterial Economy												
Publishing	120	1057	133	1251	67	368	152	918	178	1464	56	318
Retail Trade	6226	16102	3945	10082	1697	4278	5602	15026	3753	9652	1532	3998
Banks and insurance	235	6579	312	3863	162	1957	377	4398	552	5402	392	2498
Services for businesses	1112	5506	1760	6748	1109	2665	2275	10770	3484	14117	1852	5510
Public Admin.	169	10414	84	3050	67	2079	266	9071	117	4756	76	2594
Education	349	9008	337	10118	123	2992	376	8221	378	9312	151	3023
Health	547	8389	541	6876	286	3681	894	9165	937	9689	492	4109
Other services	1334	5061	1095	3573	579	1707	1544	4952	1363	3782	733	2045
Total immaterial	10092	62116	8207	45561	4090	19727	11486	62521	10762	58174	5284	24095
Rest of economy												
Industry	3050	49396	3005	23066	1505	9916	3163	38747	2637	18834	1305	8382
Other tertiary	3841	39738	4353	25084	2259	9083	4044	37792	4199	25735	1990	8766
Overall total	16983	151250	15565	93711	7854	38726	18693	139060	17598	102743	8579	41243

Source: Istat, Censimento Economico (Economic Census) 1991 e 1981, our processing.
N.B. The branch of the total industry includes mining activities, manufacturing activities (excluding publishing), energy and construction.
Other tertiary includes non-retail trade, transport, hotels and public concerns.
'Other services' correspond to section O (other public, social and personal-sector services) except for division 90.

Table 2 The immaterial economy in Venice by area in 1991

	Immaterial economy			*Rest of the economy*		*Total*		*Immaterial econ/Total.**	
	L.U.	*Emp.*	*Ave. size*	*L.U.*	*Emp.*	*L.U.*	*Emp.*	*L.U.*	*Emp.*
Historical Centre	4,495	27,165	6.0	2,584	27,895	7,079	55,060	63.5%	49.3%
Coastal Venice	768	3,724	4.8	571	3,011	1,339	6,735	57.4%	55.3%
Lagoon Venice	99	289	2.9	52	221	151	510	65.6%	56.7%
Mainland	6,048	30,356	5.0	3,633	26,098	9,681	56,454	62.5%	53.8%
Marghera	76	987	13.0	367	19,314	443	20,301	17.2%	4.9%
Total	11,486	62,521	5	7,027	76,539	18,693	139,060	61%	45%

Source: Istat, Censimento Economico (Economic Census) 1991 – our processing.
*No. of employees in whole area.

2.2. *The economic trend in the municipality of Venice during the 1980s*

A clear picture of the trends in the Venetian economy (immaterial and non-immaterial) in the 1980s can be obtained by comparing the data from the two Censuses (1981–1991). The comparison between the 1991 survey and the 1981 survey reveal a problematic scenario in terms of the trend in the Venetian economy as a whole (Table 1).

During the 1980s employment showed an overall reduction, contracting from over 150,000 employees to roughly 139,000, a total drop of 8%. This total loss of over 10,000 jobs can be almost entirely attributed to the industrial sector which registered a negative balance of more than 20% (from over 49,000 employees to 38,700). The fall in employment recorded in the industrial sector is attributable in large part to the decline of the industrial centre of Porto Marghera and to the closure of large industrial plants in the chemical and metallurgical sectors. Therefore, it should not be surprising that the balance for the local units remained positive (going from 3050 to 3163 local units). The sector which has been defined in this study as 'other services' also showed a 5% decline in number of employees, even though a growing trend is seen for the local units which is, however, insufficient for maintaining the employment levels of 1981.

The only sector which did not show signs of decline in employment terms was the immaterial economy. In terms of absolute values, employment was around 62,000, but the relative importance of this sector in the municipality's economy increases from 41% to 45% (in terms of employees). It must be stressed that the difference is due more to the decline in the material economy than to an explosion in the sector itself: in the interval between the censuses, the immaterial economy gained only 400 employees, growing by 0.7%. In any case, this fact is significant. The economic structure of the Venetian area, as a whole, is seen to evolve towards a higher level of tertiarisation. The immaterial economy proved to be the only branch which survived in the face of changes in the economic context, even though it did not succeed in obtaining opportunities for a strong growth in employment from the transformations in progress. This is further confirmed by the trend for units, which is the most sustained in absolute terms (+14%, equal to almost 1400 units).

Obviously, the complete set of data concerning the trend in the immaterial economy conceals sectorial trends which are very diversified and sometimes completely divergent. Indeed, the sectorial composition within the immaterial economy in 1981 is different from the present composition (Table 1, Figs 1 and 2). It is interesting to compare the division of the relative importance of the different sectors which make up the aggregate of the immaterial economy in order to evaluate analytically the components of the Venetian tertiary sector.

From a quantitative point of view, both in 1981 and in 1991, the retail trade was the most representative sector of the Venetian immaterial economy. In 1981, the retail trade comprised 16,102 employees, over a quarter of the total number of employees. In 1991, the sector showed a 7% reduction in employees, with a fall of about 1000 units during the period between the censuses. The local units decreased by a few more points (–10%), demonstrating a tendency towards a growth in the size of businesses. Indeed, on separating out the data according to size, it is discovered that the decrease affected the smaller enterprises exclusively (above all, those with less than three employees, in which 800 jobs are lost), while in businesses with more than 10 employees the trend is reversed. Therefore, during the period considered the city was subject to a process of rationalisation in its distribution system, a phenomenon which, moreover, characterises the evolution of a good part of the urban areas in the region (and not only), but strengthens its own specialisation in trade.

In 1981, the Public Administration sector guaranteed employment for 10,400 people. In the period between the censuses the administration suffered

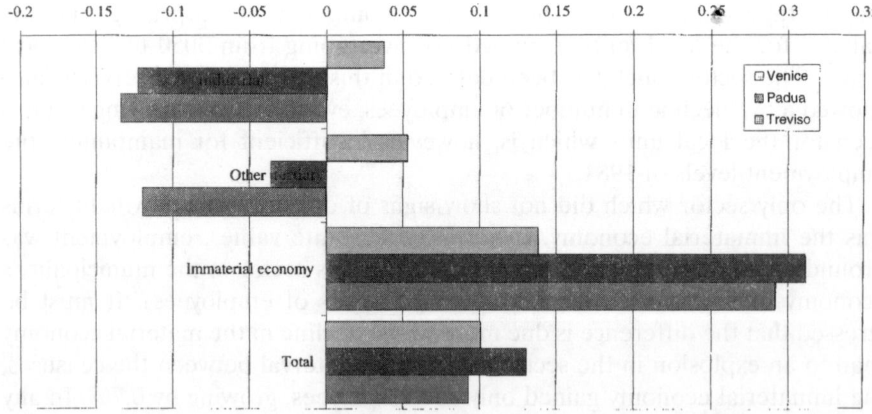

Figure 1. Local Units in the municipalities of Venice, Padua and Treviso. Variation of percentage 1991–1981

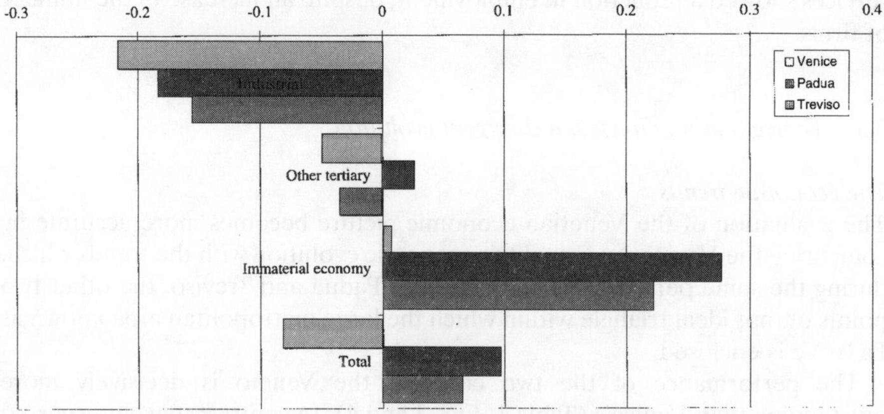

Figure 2. Employees in the municipalities of Venice, Padua and Treviso. Variation of Percentage 1991–1981

a loss of over 1000 jobs (13%) although the number of local units increased. Second in order of importance in the composition of the immaterial economy was the sector of services for businesses, the growth of which was formidable. The number of employees increases from 5506 in 1981 to 10,770 in 1991, a growth of 95.6%; the units increased to an even greater extent (from 1112 to 2275, +105%).

An opposite trend was seen in the banking and insurance sector, which lost one-third of its employees in the 1980s. The concurrent growth in units showed a sharp reduction in their average size (from 28 to 12 employees per unit). This drop must be attributed completely to a reduction in the number of large companies (over 50 employees), to which the disappearance of 3000 jobs can be attributed. This is a heavy loss which the growth in smaller companies, although it is a steady one (both in terms of units and of employees, roughly 70%), did not manage to compensate for. The figures for the insurance sector are significant since they concern one of the traditional sectors in the city's economy.

It must be remembered that part of this reduction could be attributed to a structural transformation of the sector during the 1980s and 1990s which was characterised by the breaking up of supply throughout the territory and by a consequent reduction in the average size of the individual local units. However, for a significant part, this drop must be explained as the result of strategies for locating the headquarters of credit and insurance companies which, beginning with the *Assicurazioni Generali*, have preferred to establish their head offices on the mainland, thereby giving priority to barycentric locations within the metropolitan area.

Trends in the other sectors were less significant; health was in expansion, with an increase in units (+63%), while publishing, education and the other services showed a reduction in employment, despite an increase in the number of firms.

2.3. *Venice, Padua, Treviso: a divergent evolution*

The economic trends

The evaluation of the Venetian economic picture becomes more accurate by comparing the Venice municipality's economic evolution with the trends which, during the same period, have characterised Padua and Treviso, the other two points of that ideal triangle within which the large metropolitan area known as PaTreVe is enclosed.

The performance of the two cities of the Veneto is decisively more outstanding than Venice's (Table 1, Figs 3 and 4). In contrast to what occurred in the regional capital during the 1980s, the two mainland cities experienced an increase in employment. Growth was most sustained in Padua, where it reached almost 10% (the total number of employees increased by almost 20,000 units, reaching a total of about 102,000) but growth was also seen in the municipality of Treviso where it equalled 6.5%. In both cities, industrial employment was cut, although not as sharply as it is in Venice (by 18% and 15%, respectively) while the trends in the other tertiary sector diverged, positive for Padua (+2.6%) and negative for Treviso (–5%).

The most marked differences from Venice are found in the immaterial economy. In the regional capital the immaterial economy did not expand, while in the other two cities it grew steadily: in Padua the number of employees in the immaterial sector increased by 28% (over 12,000 employed) to total over 58,000; in Treviso they increased by 22% to a total of 24,000 employed. As a result of such a high increase during the 10 years between the censuses, the immaterial sector was confirmed as the primary sector in the economy of the two cities which employs the absolute majority of workers (57% in Padua and 58% in Treviso.) In both places, the growth in employees was accompanied by a more sustained increase in local units and, as a result, by a reduction in their average size which proves to be more conspicuous than it is in the Venetian area.

In conclusion, during the 1980s, the two mainland cities underwent a process of considerable tertiarisation. Padua, which showed the most dynamic trends, reinforced its role as the tertiary centre (the principal one of the region), whilst Treviso, which in 1981 proved relatively non-specialised in service activities, makes a significant recovery. Both cities benefit from an economic growth which contributes in restoring employment in a substantial way, thereby consolidating the new sectors of the economy. In contrast, Venice, which already in 1981 begins with a figure of employees in the immaterial sector which is lower than in the other two cities, shows a limited capacity for conversion. This is largely due to its industrial inheritance which conditions the forms of entrepreneurial activities and of local industrial policies.[6]

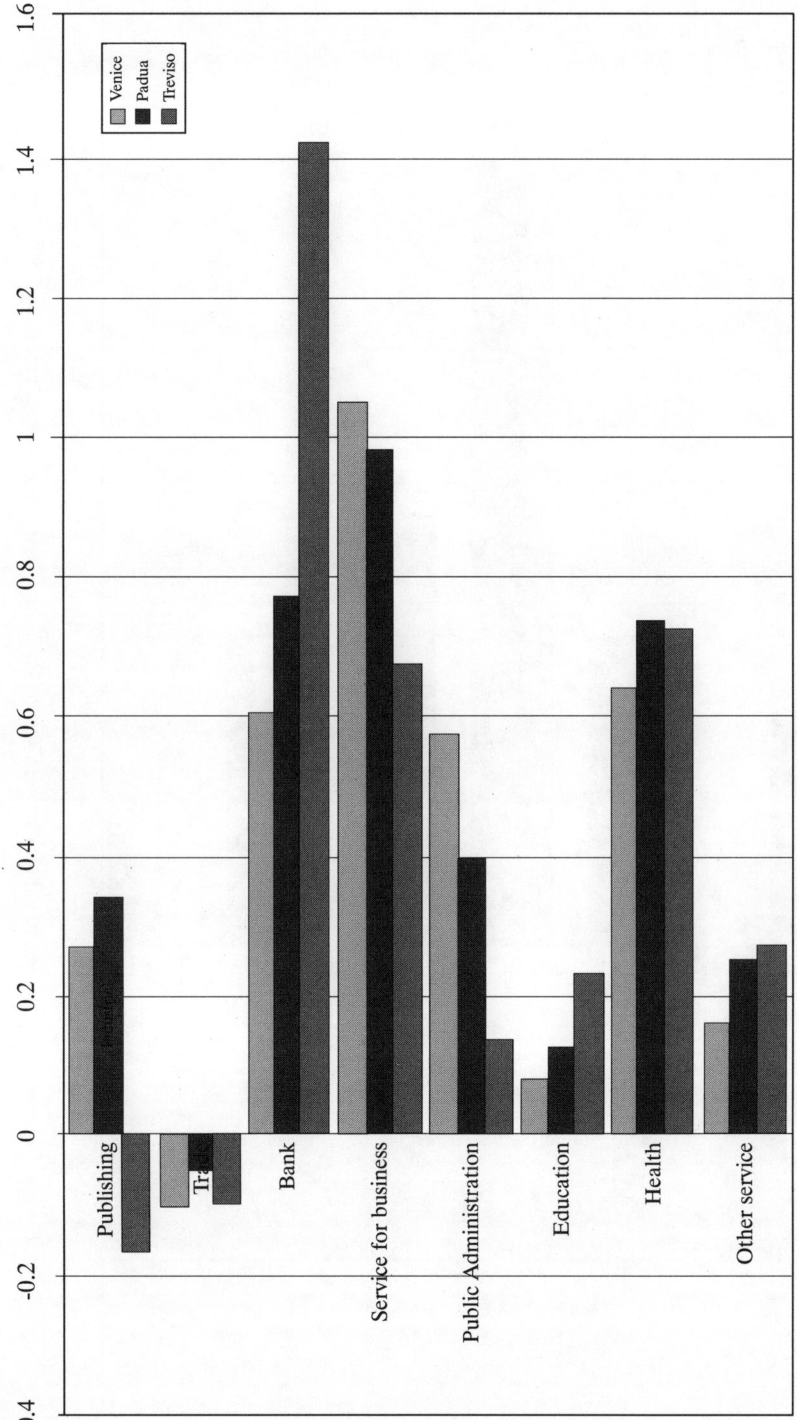

Figure 3. Local Units in the immaterial economics in Venice, Padua and Treviso. Variation of percentage 1991–1981

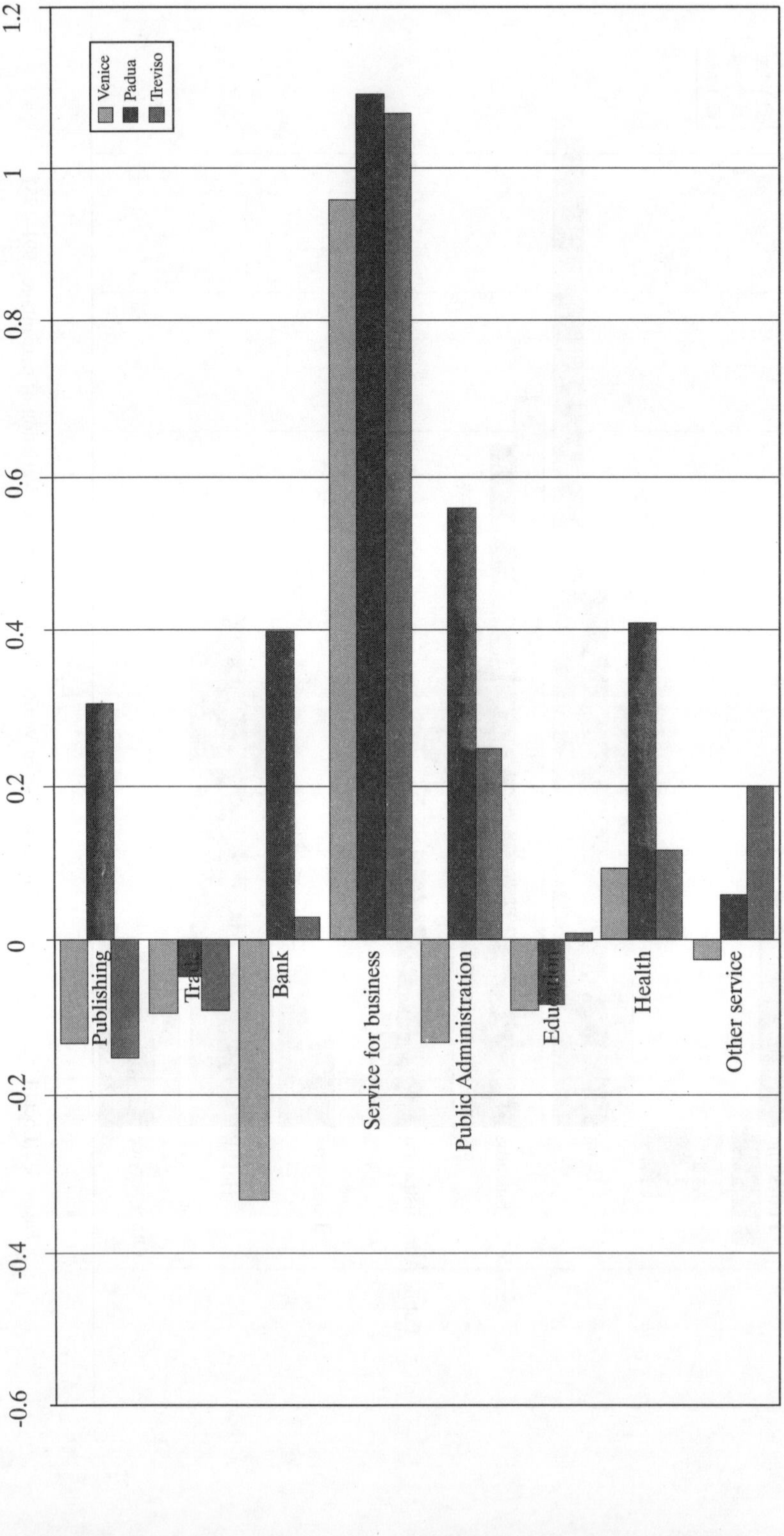

Figure 4. Employees in the immaterial economics in Venice, Padua and Treviso. Variation of percentage 1991–1981

The sectorial composition
Segregation according to sector brings highly important information to light. In order to draw a picture of the trends during the period between the censuses, it is useful to focus on the trends in various sectors (Tables 1 and 2). As far as trade is concerned, Venice's evolution takes the same form as that of the other two centres of the Veneto. During the 1990s, in the whole metropolitan area, the sector underwent an overall decline in employment and, above all, in the number of units (a decline which affects the very small units the most, i.e., those with less than three employees). In Venice and Treviso the reduction was slightly greater than in Padua (–6.5% against –5% in terms of employees) but it was accompanied by a more conspicuous process of concentration and, as a result, of rationalisation in the distribution structure. Indeed, while in Padua the decrease penalised both small and large concerns, in Treviso and, above all, in Venice, it affected the small units most of all, whereas employment increased in firms with more than 50 employees.

There were also convergent trends between the three municipalities in terms of services for businesses which perform well everywhere showing, on average, a doubling in employees. In this sector of the immaterial economy Venice has demonstrated a significant entrepreneurial vitality, with a 95% increase in employees. This was certainly an excellent result, 10 points higher than the national increase which was already extremely high itself (85%) and a little lower than the average for the northeast (97%). However, comparison of this result with those for the two cities of Padua and Treviso shows how both of these municipalities have undergone more sustained growth: Padua with an increase of 109% and Treviso with 106%. Venice remained in first place in terms of the local units' rate of expansion: +105%. In other words, the process of tertiarisation of the economy significantly affected the whole metropolitan area, including the Venice municipality, although to a slightly lesser extent. Below, we shall distinguish the impact of this growth on the different areas of the municipality, highlighting how the development of real services has very different connotations in Mestre and in the historical centre.

In the case of the Public Administration, Venice's evolution was negative compared with that of the other two cities. With an opposite trend to the regional capital (where only the units increase), in Padua and Treviso the sector expanded both in terms of employees and of units (especially in Padua where the number of employees grew by 56%). In Venice there was a sharp reduction in the average size, while in the other two provincial capitals these grew.

The sector which, in Venice, showed the worst performance is that of banking and insurance, with a sharp decline of 33% in employees. The evolution of the two mainland cities showed an opposite trend, with a considerable increase in number of employees, especially in Padua where the increase equalled 40%. Venice, which had over 6500 employees in the credit and insurance sector, a number which was far higher than numbers in the other cities (Padua had 3800 employees), lost more than 2000 jobs over the decade and surrendered the role of financial capital to Padua which, in the same period, took on 1600 employees. The sector experienced a strong development in the Treviso area as well (+26%) where, above all, the proliferation of units of a smaller size was seen.

Health expanded everywhere, particularly through a high increase in the number of units (about 70% on average). The increase in Padua was particularly marked, with 40% more employees, strengthening the city's already important hospital centre. Job cuts were made in education in both Padua and in Venice (exacerbated by the demographic drop). Only in Treviso was there a slight increase in employees and a greater diffusion of educational institutions. In Padua, the publishing sector grew, going against the general tendency.

2.4. *Some conclusions on the immaterial economy of PaTreVe*

A comparison between the various divisions of the immaterial sector in the economies of Venice, Treviso and Padua which it is possible to obtain from the census data allows various concluding evaluations to be made.

In 1981, all three cities had both a high number of employees and more units involved in trade. At the beginning of the 1980s, trade was the most significant sector of the immaterial economy in the three cities and accounted for almost a quarter of the total number of employees in the immaterial sector, Venice being distinguished by the highest values (both in absolute and percentage terms). Furthermore, the data provide a picture which is still consistent in terms of each city's traditional specialisation: public administration in Venice (with 17% of employees) and education in Padua, the city's most prominent sector, with a number of employees which is similar to the figure for trade. In 1981, Treviso showed a corresponding presence of employees in health.

By 1991, the situation had changed. As it has been seen, Venice maintained a specialisation in trade, which remained the principal sector in its immaterial economy, and had services for businesses in second place. The sectorial composition in Padua and Treviso was different from Venice's but interesting similarities between the two mainland cities are seen. In both cities, trade was replaced in first place by services for businesses, which accounted for one-quarter of the total employees. By now, the disjunction between the two sectors amounts to some thousands of jobs and, therefore, the replacement appears to have been consolidated. Trade constituted the third sector in the immaterial economy in terms of the number of employees. Education asserted its strength in Padua but was also more widespread in Treviso.

Comparing the structure and the sectorial dynamics of the three cities it is seen that, between 1981 and 1991, Venice managed to maintain its traditional specialisation in public administration (linked to its function of regional administrative capital) and in the retail trade (despite the decrease in employees, Venice continued to have the highest number of employees compared to the other two cities, both in absolute and in percentage terms). In the banking and insurance sectors, on the other hand, as a result of the reduction in employees, the city was forced to abdicate its prime position in favour of Padua which proved itself as a reputable financial centre in the area.[7] As well as the banking and insurance sector, Padua strengthened both its well-known hospital centre and its importance in the sector of services for businesses. Despite a certain reduction in employees, the city managed to maintain its traditional specialisation in education. The city of Treviso emerged, above all, for its dynamism (given its smaller size, the city cannot take first place in absolute terms). The most important result to highlight in the context of a wide-spread increase in immaterial activities is Treviso's increasing specialisation in services for businesses.[8]

An interpretation of the tendencies shown by the census data proves wrong those who have envisaged for the municipality of Venice a natural tendency towards the tertiarisation of its economy. The regional capital's history and culture have not been sufficient to distinguish the municipality as the centre of the immaterial economy, despite the fact that its tradition has been consolidated in several sectors such as banks and insurance and despite the high number of professionals. In contrast, it is precisely the capacity and innovation in industrial enterprise which has provided the principal thrust for locating the services for businesses: its establishment reflects its high interdependence with the trends of the development of the latter sector, highlighting the problematic nature of the policies for developing the Venetian historical centre. The latter, as will be made clear in the following paragraph as well, has difficulty in finding a position within the metropolitan space precisely because it remains outside the virtuous circles of development which have determined the area's success.

Therefore, the subject of Venice's vocation as the capital of the new service economy must be critically analysed in view of the data on the differentiated development of the different cities' economy. In fact, Venice's exclusion from education and innovation, the development of which has marked the growth of the central area instead, risks becoming a structural factor which could jeopardise the possibility of revitalising the municipality's and, especially, the historical centre's economy. Beneath a critical threshold, the survival of important economic subjects is at risk: to invoke a centuries-old tradition or a vocation which is, essentially, utopian does not damage a clearly consolidated development trend.

2.5. *The spatial distribution of immaterial activities in Venice*

Having sketched this general picture, we shall now move on to examine in detail the present situation in the area in question: the Venice area. What will be provided in the following paragraphs is a snapshot of the immaterial economy in the different areas into which the municipality of Venice can be divided (the historical centre, coastal Venice, lagoon Venice, the mainland and Marghera), as furnished by the data of the 1991 Istat Census (Table 2).

The immaterial activities' spatial distribution in the various areas of the municipality is by no means uniform; the great majority of units and employees (92%) are concentrated in the historical centre and in Mestre whilst only 8% of those employed remain for the other areas (Marghera, lagoon and coastal Venice).

This fact is consistent with the production and settlement features of the three areas. In the case of the islands and the coastline, it is explained by the small number of total employees. In actual fact, these two areas are by no means lacking in tertiary activities. They simply have few workers employed, both in the sector of the immaterial economy and in the other sectors. In the case of Marghera, on the other hand, the data shows a marked tertiary despecialisation – a good 95% of those employed in Marghera work in 'non-immaterial' activities – and this appears to correspond logically to the typically industrial nature of the area (which only contains the port and production works area). As a result, to speak of the immaterial economy in the municipality of Venice means, essentially, to speak of the immaterial economy in Mestre (mainland) and in the historical centre: these will be the two territorial entities which we shall focus on.

Moving on to examine the sectorial distribution of the employees within the different areas, it is noted that, with the exception of Marghera, the immaterial represents around 50% on average. In Mestre, those employed in the immaterial sector represent 54% of total employees, whilst in the historical centre they remain just below a half (49%). On observing the local units, the positions are reversed: the historical centre overtakes the mainland in terms of the number of units (63.5% against 62.5%) and in terms of their average size (six employees against five). On the coast, the units of the immaterial sector make up 57% of the total units and the employees form 55% (average size is 4.8). In lagoon Venice, there is a higher number of units (66%) and their average size proves to be decisively smaller (2.9).

2.6. The sectorial composition in the different areas of the municipality of Venice

We shall now focus further on the situation by subdividing the branch of the immaterial economy so as to examine the individual sectors which make it up. We can see each sector's distribution of employees and units in the various areas of the municipality (Table 2).

The situation in the three minor areas is easily described. The quantitative compositions are on average low (obviously, in relation to the low number of total employees and of resident inhabitants), as are the percentage values. Lagoon Venice's role in the whole of the Venetian economy appears negligible considering its very slight impact on the totals of the sector. The coastal area is more visible (between 1.1 and 7%) and shows a fairly important position in health (14%). In Marghera, on the other hand, a significant amount of services for businesses is seen (6.6%).

The situation in the historical centre and the mainland is more complex. As far as the quantitative composition is concerned (expressed in terms of employees), some similarities are seen, as is an important difference. First of all, it is shown that, both on the mainland and in the historical centre, the sector which counts the highest number of employees is the retail trade. In the retail trade, 52% of the sector's employees work on the mainland; the historical centre must yield its position, although it has a high percentage (40%). The units are respectively 47% and 44%. In the centre, the large majority of workers (87%) is employed in concerns with fewer than 10 employees. On the mainland, this figure drops to 71% and the larger firms become more dense.

The main difference between the mainland area and the historical centre can be seen in considering the sector which occupies second place in order of importance in terms of employees: Mestre is characterised by its more than 6700 employed in services for businesses; the historical centre, on the other hand, is distinguished by its 6000 employees in the Public Administration.

Over 60% of the units and of the employees in services for businesses are located on the mainland, with roughly double the number of the units and employees of the historical city (35% and 28% respectively). Furthermore, on the mainland the units are larger on average. The island of Venice makes its role of historical centre of the regional capital city carry weight by amassing 67% of the total number of employees in the public administration and over half of the units, which show average sizes which are definitively larger (45 employees; the figures for Mestre are 31% and 25 employees).

This result confirms the two points already raised above. The first concerns the historical centre's integration within the metropolitan area of Venice-Padua-Treviso: the concentration of services for businesses in the mainland area confirms the growing separation between the historical centre and the trends in economic development which govern the growth of the Northeast. Mestre benefits from the existence of the tertiary sector with higher value added without involving the historical centre, which is isolated by a system of infrastructure for connections which make it practically inaccessible. The second aspect concerns the historical centre's traditional administrative vocation: both the existence of the local administration and the location of the University are subject to criteria which are very different from those which characterise the location choice of businesses. However, that said, it must be emphasised that also these institutions, although within greater limits, demonstrate that they are quite attracted to the areas of larger growth and they will not necessarily be able to compensate for a distance from the centre (or from the centres) of economic growth as a result of a territorial specialisation design of a purely political nature.

Mestre also eclipses the historical centre in terms of employees in the education sector (49%) and the health sector (46%); in this case, it is a matter of a relative majority, not too far from the percentage of island Venice (43% and 39%) and consistent with the public service nature (and, therefore, wide-spread over the territory for serving the population) of the two types of service (indeed, the coastal area also becomes significant, going over 7%).[9]

The historical centre also manages to maintain a predominant position in the case of banks and insurance companies both in terms of employees (both in absolute and percentage terms) and in average size, in honour of its past specialisation; the figures are 51% against 46%; the average number of employees is 17 against 9.5. On the other hand, Mestre has a higher number of units (57%).

In conclusion, the data analysis demonstrates that immaterial production is distributed between the two main areas of the city in a differentiated manner. The examination of the distribution of employees, which is the principal indicator available, highlights the present specialisation of the two areas. The conclusion which can be reached is that, alongside the city's traditional vocation sectors which are confirmed (for example, activities linked to the Public Administration in the historical centre), there are others sectors which shrink (insurance companies). On the other hand, however, the emergence of new specialisation is recorded (such as the services for businesses on the mainland) or the evolution of pre-existing specialisation is seen (the large retail parks in Mestre in place of the traditional retail trade). As a whole, a problematic picture is drawn for the development of an immaterial economy in the historical centre: in comparison to a highly dynamic metropolitan area during the 1980s which was able to revitalise its structure of services in support of the industrial firms, the area of Mestre seems to be essentially integrated, whilst the historical centre seems to hide behind vocations which are not sustained on the basis of market theory. From the immaterial economy perspective, PaTreVe appears more like PaTreMe, or, rather, Padua-Treviso-Mestre: today the Venetian historical centre suffers the consequences of the difference in physical accessibility which does not allow forms of economic location. Despite the declarations and the hopes, the immaterial economy at higher value added does not seem to have taken root within island Venice.

2.7. Concluding reflections on the current trends of the Venetian immaterial economy

The analysis of the Venetian immaterial economy built on the evaluation of the census data produces a picture of an economic situation which is difficult to sustain: the economic profile outlined above highlights the historical centre's increasing difficulties in maintaining its own social base and in finding a new economic equilibrium by means of an upgrading not exclusively based on the tourist industry. The aspect which deserves further emphasis is the increasing separation between the features of the economic development of the historical centre from Mestre's: in terms of the immaterial economy, the decade of the 1980s heightens the separation between the two parts of the city, confirming island Venice's incapability to benefit from the economic development which characterises the central Veneto area as a whole. During the decade of growth, the historical centre has visible difficulty in keeping up with the mainland even in those sectors such as finance and insurance, for example, in which the city boasts a tradition and prime position and which have their roots in the city's history.

The alarming fact which emerges from the census data is the weakness with which the historical centre competes for the location of the new immaterial economy. To imagine Venice as the natural centre for the establishment of services of high value added such as business consultancy or software production, for example, appears to be extremely unlikely in light of the trend in locating and in the growth in services to businesses: the services are not located on the basis of criteria which prize generic centres of prestige, but grow and consolidate their presence in those areas which demonstrate a higher capacity for growth and development.

If we then consider the more traditional kind of activities, the trend data does not appear any more reassuring. The analysis of commercial activities shows a traditional type of structure for the sector, in which the signs of a rationalisation process are still barely visible. Compared to the situations in Padua and Treviso, the relative importance of the retail trade in the broader the immaterial economy of the different cities is higher, and the structure of the sector still appears extremely fragmented.. Furthermore, it must also be remembered how, in the face of a drop in the resident population of about 1,000 units in the space of time examined, the retail trade maintains the number of local units unaltered which suggests a progressive reorientation of these units from the local resident market to the tourist market.

Furthermore, the census data for 1991 cannot take into account the extraordinary transformation undergone by trade following the arrival of the large national and international commercial operators which have focused on the opening of commercial centres and of innovative sales structures for the Italian context, such as hard discount. In the space of a few years, the Mestre hinterland has become populated by proper commercial centres which are added to the traditional commercial centre of the Romea district. These structures have caused a deep change in the sector's equilibrium, triggering a crisis in small-scale supply not only in the Venetian context, but also in that of Mestre.

On the other hand, the confirmation of the historical city's traditional leading position in terms of the presence of public administration and of the university should not be surprising: in both cases, the city's specialisation compared to the territorial system is the result of a well-defined political decision which has given priority to the historical centre in the knowledge that, in a good number of cases, the public operator is the only subject which is able to shoulder the expenses entailed in the ordinary and extraordinary maintenance of the prestigious buildings in Venice. What is very striking is the poor capacity of these same structures to become the draw for economic initiatives of even a limited size which could effectively exploit the external economies which are generated by the presence of the educational institutions. This is particularly true for the university which, despite the vitality of the policies for development and diversification in the supply which have been made over the last few years, still does not seem to be a real point of reference for the city's immaterial economy.

Therefore, the scenario outlined confirms the hypothesis of a Venetian specificity which affects the system of economic specialisations which characterise the historical centre. Venice's insularity constitutes the prime cause of a process which tends to remove Venice from the process of the Northeast's economic development, bringing not only its metropolitan integration into question as much as, rather, the very idea of Venice as a city which is able to maintain its own economic equilibrium independently of a system of subsidies determined on a political basis.

In a scenario which we have defined above as an inert one, in which policy is limited to recording the social and economic trends currently in progress, the problem of sustainability appears important. If sustainability is at the same time economic equilibrium and the maintenance of a local structure which can preserve, also in evolution, a city identity, the current economic trend can hardly be considered sustainable, at least within the present institutional framework. As has already been mentioned above, the growing discrepancy between the islands and mainland is a divisive factor between the historical centre and Mestre. It poses completely different problems for governing the city (for example, policies on tariffs on public services or policies on taxing homes) and pushes opposing forms of development (the revenue culture against the entrepreneurial culture), thereby providing the structural premises for an institutional separation between the two main areas which make up the municipality of Venice.

To imagine a different future, we must consider the economic prospects of those sectors which can counterbalance the impact of the tourist sector in economic and social terms. In the following, we will try to emphasise how the information technology networks can impact the present structures of supply. If in one respect the network offers an extraordinary opportunity for upgrading these two branches of activity, not only in terms of a reduction in costs and operational efficiency but also in terms of a comprehensive reflection on their own activities, in another respect, the new technology could exacerbate, even more markedly, the imbalances which already characterise the city's economy today.

3. The expected transformations and the possible policies

3.1. Post-Fordist production and the city's evolution

The evolution of Venice's urban system must be placed in the context of a change of paramount significance, with the transition from the Fordist to the post-Fordist paradigm.

After half a century of Fordism, the city's economy is already today distinguished by an absolute predominance of immaterial production: in the face of a material production (manufacturing) which, since it needs large spaces, has progressively chosen the countryside, the cities, instead, have specialised in immaterial production. Indeed, it is the immaterial services which constitute the most significant product in absolute terms provided by the urban centres; and the business activities of research, management, communication which are concentrated in the headquarters are immaterial. Furthermore, these headquarters are, turn, located in the urban centres.

Nevertheless, the immaterial character created by Fordism does not show the explosive features which are witnessed with the advent of new information technology. Indeed, the immaterial resources which typify the Fordist system are given by knowledge and relationships which remain incorporated in a material base from which it is difficult to separate them. They are incorporated in the goods which are exchanged on the market, in the intermediate products which pass along the production line, in the machines which spread the technological innovations, in the infrastructure which equips the territory, in the mental models employed by people, in the organisational culture. As a result of their rooting in the materiality of a base which is not easy to eliminate, the immaterial resources employed by Fordism prove to be barely transferable spatially and they tend to remain inside the individual systems (businesses) and on the territory (districts, cities). With these characteristics, they can be controlled and accumulated only with limited differences compared to the material resources.

With post-Fordism, things are profoundly changed. For two reasons:

(1) The immaterial resources, due to the rapidity and reliability of long-distance interactive communication, are increasingly transmitted in a digital form, and they tend, therefore, to be separated from their material support;
(2) Even when the immaterial resources remain bound to a specific context, the new information and communication technology make forms of sharing of experience contexts possible even between people and organisations which are a long way from each other.

Through these two complementary elements (mobility of the codified information, trans-territorial expansion of the realms of experience) the post-Fordist economy attains a regime of knowledge use which is radically different from that which has prevailed until now within the Fordist organisation of production.

3.2. Venice's present specificity: contraction of the networks used and decline in the city's vitality

The prospect of the post-Fordist de-materialisation takes on a very particular significance in Venice. This can be seen if we look retrospectively at the way in which Venice's specificity has taken shape.

Within the circuit of Fordist production, with the characteristic feature of being material, Venice, because of its characteristics, represents the exception which is destined to remain in the margins, outside the rule. Indeed, the city cultivates a diversity which separates it from the supporting axis of innovation and generation of value. It is no coincidence that the economic boom of the Northeast has taken place with mechanisms and with subjects which have nothing to do with Venice, the old capital of this region.

The Venice of today has actually become, in the context of material production, a city which is radically different from the city-capital which history speaks of. The city-capital was cosmopolitan, open to inter-cultural influence between different peoples, projected great distances by a logistic network (via the sea) which was capable of sustaining a highly respectable international division of labour at least until the end of last century. In our century, mass production and automobile transport have progressively limited the Venetian economy. With the loss of the capacity to compete in the field of manufacturing logistics and of trade logistics (like many other sea cities), the long-distance network which sustained the financial power of Venice-capital has been drastically curtailed, to the point of becoming a local network which is really insular. Its system of relationships has contracted, the division of labour, if tourism is excluded, has become increasingly limited to the intra-urban circuit, economies of scale have been lost. Thus, the historical city has ended up cultivating a state of uniqueness (the city-island, the city without cars which does without the network of connections which is guaranteed by land transport) which, structurally, is associated to a state of slow economic, social decline as well as a decline in its innate characteristics. What has not suffered from remaining outside the wide networks is tourism for which the historical centre's separation from the mainland activities, if anything, forms a factor of attraction, a valuable feature of diversity which is appreciated from the perspective of the tourist visit. Nor have the activities of providing services for the island market suffered from it: the retail trade, services for residence and housing, local transport. These are all activities which, in a certain way, have been protected by the boundary caused by the lagoon and which have, also today, based on this boundary their specialisation and their competitive advantage compared to the potential competitors of the mainland.

The passage from the long networks of the Venice-capital to the short networks of island Venice has favoured specialist activities bound to the city's uniqueness in terms of the lagoon and of tourism but these have only been partially able to replace the sustaining specialisations of the old city-capital (trade, finance, the port, marine activities public administration), which have been condemned to a slow but inevitable decline.

3.3. The new possibilities of post-Fordism: Venice can reconquer the long networks

The trend towards the post-Fordist de-materialisation of production changes the cards on the table. Furthermore, it creates a situation which is completely new and which certainly holds further threats to sustainability but which offers the city many new opportunities as well. First of all, in production of a strong immaterial content, the speciality of Venice bound to its insular nature, that is, logistically separated from the mainland circuits, is lessened. For the communication networks which convey digital information, rather than material goods, the boundary caused by the lagoon is simply irrelevant. From this point of view, Venice has the possibility again of being an active point (by now, no longer a centre) in long networks, on a metropolitan or global scale.

The post-Fordist de-materialisation of production does not, in itself, bring either advantages or disadvantages to the city. However, it does set the situation back in motion, by opening up new prospects and eroding previous certainties. Venice's role will emerge from a competition between cities which, in the process of a general reallocation of functions and activities, offer different possibilities and resources. However, it will also be the result of co-operation, that is from the intentional construction of networks of relationships and of the division of labour which each city will be able to use for competing.

Co-operation and competition will be able to determine a fundamental change in the present allocation of production activities and of residence, intervening in a situation which is in motion on three different levels:

(1) New immaterial production activities of a high value added and of a low environmental impact are developing which usually require little logistic support which is compatible with Venice's lagoon and water specificity. One part of modernity can, at least potentially, be united in Venice, overcoming the boundary caused by the lagoon.

(2) Existing production systems are becoming modulated and, consequently, they change their location geography. As a consequence of the new communication and automation technology, it becomes advantageous to break up the long, vertical and sequential chains which characterised the Fordist production and to reallocate the individual stages and modules of production throughout the territory. This tendency liberates an increasing amount of businesses, of capital, and of activities from traditional locating and makes the towns and cities compete with each other for intercepting the production sectors which are susceptible to reallocation. Venice's position in this global reconfiguration of location depends, therefore, less on the protection provided by the boundary caused by the lagoon and more on how much the city has to offer the different activities;

(3) Places of residence, of study and of work tend, thanks to the support of the long-distance interactive communication network, to become separated from the location of firms and thus acquire an autonomy which makes it possible to increase the influence of location advantages which are specifically aimed at study, work, and residence. It is a matter of a disassembling and reallocating effect which has the same theory as modularisation (point 2), but which affects individuals. Potentially, every person can be considered an autonomous module from the locating point

> of view. Therefore, competition between different places could take place by taking advantage of the specific demands relating to study, residence or work. Venice, which has specific features for residence which are highly positive in some respects, and highly negative in other, can be affected in a crucial way by the process of making location autonomous which, in the city of immaterial production, affects residence, study and work.

If the importance of the above-mentioned reallocation phenomena and of their specific effect on Venice are taken into consideration, something becomes clear: i.e., that what the next century has in store is no longer the slow and dignified decline of the historical city, but a period of accelerated evolution in which the reasons for leaving Venice could be multiplied but, at the same time, so could the reasons for returning there or for choosing it from the many possible cities.

In fact, the new city of immaterial production will be the result of the balance (between entries and exits) relative to the three distinct reallocating processes (one for each of the points examined above):

(1) redistribution throughout the territory of the new activities;
(2) redistribution of the modules 'freed' from the existing activities;
(3) redistribution of places of residence, study and long-distance work.

3.4. *Attracting new activities*

As far as the first point is concerned (attracting new activities), the process of competition, co-operation required for obtaining them is closely tied to two key factors. Firstly, the time required for the process of installing the infrastructure for the territory and secondly, the rates of developing specialist skills which it will be possible to generate in the Venetian area in view of its upgrading. Judgement on both the timescales and the skills must relate to how much the potential competitors will do: if the decision-making processes for Venice and the disposition to experiment in new fields both prove to be lesser than those of other cities, it is possible that the process, however well intentioned, will end up with limited results.

In the spontaneous evolution of things, it is likely that Venice will have a certain capacity for attraction in the sectors of immaterial production which, whilst adopting new kinds of techniques and offering innovative products, originate from pre-existing ones. Undoubtedly, culture and art, to make two significant examples, are destined not only to grow as a proportion of the total demand over the next few years, but also to take on radically new forms of production and utilisation. A revitalised supply of cultural and museum services does not simply depend on the availability of innovative technology. A revitalised cultural supply begins from what already exists and from the historical inheritance of which Venice is still a legitimate repository today.

3.5. Redistributing what exists

In its capacity to develop or to attract new immaterial businesses, Venice has a high possibility for success, even if it requires a powerful injection of technological investments and the starting of experimentation which are both currently absent or are only in the starting stages.

Venice's reunion with the mainland via networks will be, at least to begin with, a further crisis factor for those activities which are made more vulnerable by the metropolitan competition and a further cause of functional delocalisation. Indeed, residents in Venice will be able to turn to service suppliers in the metropolitan area who can be reached with relative ease via the network. The suppliers, in turn, will be able to organise themselves for contacting their potential Venetian clients in an interactive and personalised way. The model is applied to an extensive range of services which goes from trade to health.

Nevertheless, in the redistribution of functions on a metropolitan scale, Venice also has relatively strong positions which can attract demand from the territory. In this case as well, besides tourism, it is possible to think of a specialisation on a metropolitan scale in cultural, museum and historical–artistic functions. In these fields, it is not only a question of developing new sectors, but also of redistributing the existing market, by relying on the greater mobility of supply and demand.

3.6. A greater degree of liberty in the residence, study and work relationship

Finally, it is necessary to consider the redistribution of location for residence, for study and for work by means of the networks. The main effect of the digital revolution is that of making the choice of the place of residence increasingly independent of professional commitments of study or of work, which could be carried out in the form of tele-education or tele-work. This makes it possible to live in Venice even if one works (or studies) in another city, perhaps a long way away. Or, on the other hand, it makes it possible to move out of Venice even if one works or studies in Venice. In this way, the choice of where to live becomes largely independent and, therefore, is based on specific factors.

This is an opportunity which must be carefully considered and which could, if adequately supported, have large effects on the city's future. Indeed, the multiplier effects, which are precisely those of the global networks, apply even to residential movements: if some triggers are added to the usual reasons of attraction (historical–artistic and natural), it is not difficult to envisage that the demand for new residence in the city could take off and go beyond the hundreds of aspiring residents (from the whole world) to reach thousands of them. It is certainly necessary to create the logistic, building and communication conditions so that this latent demand can be fully revealed in future years.

Table 3 Local units and employees in the immaterial economy in Venice by sector and by area

	Historical Centre			*Coastal Venice*			*Lagoon Venice*			*Mainland*			*Marghera*			*Total*		
	L.U.	*Emp.*	*Ave. size*	*L.U.*	*Emp.*	*Ave. size*	*L.U.*	*Emp.*	*Ave. size*	*L.U.*	*Emp.*	*Ave. size*	*L.U.*	*Emp.*	*Ave. size*	*L.U.*	*Emp.*	*Ave. size*
Publishing	75	340	4.5	4	12	3.0	0	0	0.0	72	559	7.8	1	7	7.0	152	918	6.0
Retail Trade	2445	6062	2.5	439	990	2.3	66	156	2.4	2638	7759	2.9	14	59	4.2	5602	15026	2.7
Banks and insurance	129	2233	17.3	21	84	4.0	1	5	5.0	215	2047	9.5	11	29	2.6	377	4398	11.7
Services for businesses	797	2989	3.8	77	296	3.8	2	2	1.0	1364	6772	5.0	35	711	20.3	2275	10770	4.7
Public Admin.	135	6049	44.8	12	110	9.2	3	11	3.7	111	2795	25.2	5	106	21.2	266	9071	34.1
Education	127	3503	27.6	29	595	20.5	6	79	13.2	212	4021	19.0	2	23	11.5	376	8221	21.9
Health	266	3597	13.5	66	1301	19.7	7	11	1.6	555	4256	7.7	0	0	0.0	894	9165	10.3
Other services	521	2392	4.6	120	336	2.8	14	25	1.8	881	2147	2.4	8	52	6.5	1544	4952	3.2
Total	4495	27165	6.0	768	3724	4.8	99	289	2.9	6048	30356	5.0	76	987	13.0	11486	62521	5.4

Source: Our processing of data from Istat, Censimento Economico (Economic Census) 1991

Notes

1. Except where otherwise specified, the term 'Venice' means island Venice, i.e., the historical centre with Murano and Burano.
2. The abbreviation PaTreVe (Padua, Treviso, Venice) indicates the metropolitan intersection of the three cities which are presently in the process of integration and reciprocal specialisation, as parts of a unitary metropolitan space (Costa, 1991).
3. Our thanks go to Paola Guerra and Fabio Occari for the contribution provided for this part of the chapter.
4. The data we shall use is that obtained from the two Economic Censuses of 1981 and 1991; it relates to employees and local units both in businesses and in institutions. The data will be supplemented by some data on the population which can be taken from the 1991 Population Census.
5. The term immaterial production is used to signify a part of the branch of production which is traditionally defined as services. It concerns the following sections or divisions in the ATECO 91 classification:
   - publishing (division 22), traditionally included in industry;
   - the retail trade (division 52);
   - banks and insurance (section J);
   - the activities of real estate, rental, information technology, research and other professional and entrepreneurial activities, which can be summarised in the division 'services for businesses' (section K);
   - Public Administration (section L);
   - Education (section M);
   - Health (section N);
   - other public, social and personal-sector services (section O, except for division 90).

   The tourism sector has been deliberately excluded since it is the subject of a study of another working group in the Fondazione Mattei.
6. However, in 1991, the proportion of immaterial units out of the total number of units is similar in all three municipalities: 61%.
7. Padua's role of a centre for savings is also confirmed by other indicators, including those concerning more recent years. Indeed, the Province of Padua is confirmed as the province of the Veneto with the highest absolute value of bank deposits in 1995 as well.
8. For example, focusing on one of the most dynamic sectors (banking), it is interesting to note that, during the 1990s, the Province of Treviso overtakes the Province of Padua in the classification of banking employees, thereby qualifying in 1995 as the province of the Veneto with the highest employees:deposits ratio and, therefore, as an economic structure which is an importer of capital.
9. The data shown refers to employees. In terms of units, on the mainland these represent an even larger portion of the total whilst their average sizes are smaller.

Bibliography

Becattini, G. & Rullani, E. (1993). Sistema locale e mercato globale. *Economia e Politica Industriale*, 80.

Costa P. (1991). Area metropolitana centro-veneta e città metropolitana di Venezia. In: Costa, P. (Ed.) (1993). *Venezia. Economia e Analisi Urbana*. Milan: Etas Libri.

Dematté, C. (1995). Le autostrade elettroniche cambieranno il modo di vivere e l'intera economia. *Economia e Management*, 2.

Di Bernardo, B. & Rullani, E. (1990). *Il Management e Le Macchine. Teoria Evolutiva dell'Impresa*. Bologna: Il Mulino.

Micelli, S. (1998). Il consumo post fordista. In: Rullani E. & Romano L. (Eds.) *Il Post-Fordismo.* Milan: Etas Libri.

Reinghold, H. (1995). *Comunità Virtuali*. Milan: Sperling and Kupfer.

Rullani, E. (1994). Dematerializzazione. In: Caselli, L. (Ed.) *Le Parole dell'Impresa. Guida alla Lettura del Cambiamento*. Rome: Eni-Isvet.

Rullani, E. (1995). La politica del post-fordismo. *Il Ponte*, 4.

CHAPTER 8

Governing the Sustainable Development of Venice: Elements of the Institutional Planning Procedure

Bruno Dente, Cinzia Griggio, Andrea Mariotto, Carolina Pacchi

1. Introduction: study objectives

The objective of this chapter is to provide both the necessary analytical elements and initial proposals for the establishment of a Local Agenda 21 process in Venice. Indeed, to plan and govern a process of sustainable development in the regional capital on the lagoon has several specific and unique features, just as the city's characteristics are specific and unique. In particular, and without anticipating what we will say in the pages which follow, the key elements of the problem are:

- the exceptional importance of the historical centre's architectural heritage and the consequent protection requirements;
- the lagoon's natural value which must be protected against the dangers of pollution and erosion and which must be potentially enhanced;
- Porto Marghera's industrial decline which presents the problem of changing the development model;
- the need to at least stabilise the exploitation of tourism, in order to avoid the risks of a monoculture which is unsustainable in social terms;
- the issue of protection against hydrological risk in the context of the global terms of the global climate change;
- the national and international importance of the lagoon city and the consequent jurisdictional overlaps, within a process for transforming the structure of the State in Italy.

In essence, the problem tackled concerns the features which must be adopted by the long-term plan construction process which might compensate for the unacceptable nature, in social, economic and environmental terms, of the

I. Musu (ed.), Sustainable Venice, 227–261.

spontaneous evolutionary trends and which could set the modification of such trends as an objective.

2. Local Agenda 21 experiences

2.1. The implementation of Local Agenda 21 in the world

The event which led to this study was the movement which goes under the name of Local Agenda 21. Therefore, the first questions we asked concerned the objectives pursued, the processes started and the consequent results in Local Agenda 21 experiences in Italy and, especially, abroad.

This process is based upon two fundamental elements: (1) the definition and implementation of a programme for environmental action in the long term which could establish objectives and concrete action plans at a local level; (2) the creation of consensus and of dialogue between citizens, local associations, business and public authorities. In fact, consultation of the local community aims to identify its needs, to establish the resources which each party can provide, and to identify and prepare for the potential conflicts between different interests.

An in-depth analysis of concrete Local Agenda 21 initiatives reveals a considerable variety and flexibility in both the approaches and in the instruments employed. There does not exist, therefore, a standard methodology, which is consistently applicable in different contexts, since every situation, having peculiarities and specific requirements of its own, will have to construct its own procedure towards sustainability in its own particular way. Nevertheless, some elements do exist which are almost always found in the Local Agenda 21 processes and which cannot be ignored if one wants to begin an initiative such as the one in question.

The first issue which should be dealt with is the search for and creation of forms of collaboration (partnership) between the various actors who could play an active and decisive role in the project. This usually happens by means of setting up a Group or Civic Forum which is made up of all the actors who are important at the local level and which aims to define and implement a strategy for sustainable development; defining the purpose of the process, its geographic reference area, the timescale, the resources necessary and the means of financing; the creation of Specialist Working Groups, usually made up of fifteen or twenty experts in particular subject areas who have the task of analysing the problems, of carrying out technical assessments and of proposing possible solutions.

The result of this first stage should be the establishment of a long-term, shared vision for the sustainable development of the local community. Indeed, the adoption of a theoretical approach and of a common philosophy should guarantee the identification of coherent strategies which can combine environment and development.

The second phase involves the identification of the problems and an accurate analysis of their causes. To do this, information and data on both the city's environmental conditions and its socio-economic situation is necessary. A continuous monitoring of the natural and the socio-economic environment could also prove very useful. This monitoring could be carried out through defining and developing a set of sustainability indicators created ad hoc with the collaboration of local experts and through consultation of the community. This phase can lead to the preparation and publication of a report on the state of the environment which could reveal the features and the problematic aspects of the area under examination. The identification of the problems and the establishment of a scale of priorities should take place, therefore, by means of both consultation of the community and the adoption of instruments for participation and communication such as focus groups, meetings, workshops, conferences, interviews with the local population, etc. Alongside these kinds of instruments, the consultation and support of a panel of experts who can employ technical methods for evaluating and establishing priorities should also be provided for.

The preceding phases, though very important, form, however, only the preparatory work for creating a attainable and effective action plan. The definition of objectives, of specific targets, of strategies and of possible solutions thus constitutes the crucial stage in the whole Local Agenda 21 process.

The formalisation of the plan and its subsequent adoption by the local authority will be followed by a phase in which the document itself is presented and disseminated so that the population and all interested actors become involved in implementing the strategies. The plan's implementation will have to be evaluated and monitored. The feedback process will make it possible, in fact, to invest the action plan with all the necessary modifications for improving the entire Local Agenda 21 process.

2.2. *International Experiences of Local Agenda 21*

In order to make an in-depth analysis of Local Agenda 21 experiences at the international level we have considered a sample made up of the following areas: Hamilton-Wentworth, Seattle, London, Lancashire County Council, Cheshire County Council, Leicester City Council, the Hague, Kanawaga Prefecture and Troyan. Table 1 groups the experiences analysed and shows the features of the Local Agenda 21 process started; for example, the promoter of the initiative, the actors involved, the instruments used, the issues and the priorities confronted, the results obtained.

As it can be seen, all the areas considered have serious problems of pollution and in managing natural resources. Local Agenda 21, therefore, is almost always seen as a tool for dealing with problems of an environmental nature. The process is generally set in motion by the local authority, following a top-down approach, even if the participation of all social groups is deemed a

Table 1 Local Agenda 21 experiences

City *Process Characteristics*	*Hamilton-Wentworh, Canada*	*Seattle, USA*	*London, UK*	*Lancashire County Council, UK*	*Cheshire County Council, UK*
Information about the city	450,000 inhabitants, strongly industrialised, air and water-pollution problems	500,000 inhabitants, economy based on trade and the textile, mechanical and iron industry	7 million inhabitants, problems concerning rubbish disposal, car traffic, unemployment	County of 1 million inhabitants, strongly industrialised, considerable environmental impacts	County of 1 million inhabitants with environmental problems related to the presence of industrial activities
Start date of the local A21 process	1989	1989	1995	1989	1995
Promoter of the initiative	Local Authorities	Mayor	ALG (Association of London Government)	County Authorities	County Authorities
Instruments for Participation	Civic Forum, focus groups, specialist working groups, workshop, conferences, press and television coverage	Civic Forum, Technical Committees, mass media, conferences, distribution of questionnaires	Discussion groups experts panels, open meetings, surveys	Information leaflet distribution, working groups, information in schools	50,000 information leaflets distributed, specialist working groups
Number of actors involved	More than 1000 citizens and 160 local associations	70% of the population has been consulted, 15% has actively participated	1000 people interviewed		1000 people
Issues dealt with	Natural areas, air, water, rubbish, energy consumption, land-use, transport, public health, agriculture, economy	Rubbish, air pollution, unemployment	Civic commitment, economic vitality, energy consumption, public health, mobility and accessibility	Air, energy, transport, noise, water, rubbish, territory, nature and open spaces, environ-mental education	Energy, territory, water consumption, air naturalistic issues, transports, rubbish, participation
Documents drawn up	Document on sustainability scenarios, Action Plan, Set of indicators	Long term Action Plan (deadline 2012) Set of indicators	Agenda 21 Report for London Set of indicators	State of the Environment Report Action Program	Environmental Action Plan
Established objectives and targets	Environmental, social, economic objectives for the short, medium and long term	Short, medium and long term environmental, social, economic	Not yet defined	Medium and long term environmental objectives	Short and long term objectives, mainly environmental ones
Process aims	Combine development and environmental protection in a long term perspective	Combine development and environmental protection in a long term perspective	Change in the population lifestyle through an involvement strategy	Dissemination of knowledge on environmental matters, making citizens aware	Dissemination of knowledge on environmental matters, making citizens aware

Continued …

Table 1 Continued

City *Process Characteristics*	*Leicester City Council,* *UK*	*The Hague,* *Netherlands*	*Kanagawa Prefecture,* *Japan*	*Troyan,* *Bulgaria*
Information about the city	280,000 inhabitants, economy based on the tanning, chemical and textile industry	450,000 inhabitants, service industry based economy	County of 8 million inhabitants, strongly industrialised with several environmental problems	46,000 inhabitants, severe air and water environmental problems
Start date of the local A21 process	1994	1994	1992	1992
Promoter of the initiative	Local Authorities	Local Authorities	County Authorities	ISC (Institute for Sustainable Communities)
Instruments for Participation	Specialist working groups, public meetings, questionnaire distribution	Specialist groups and information leaflet distribution, mass media, conferences	City Committee, Enterprise Committees, public meetings, mass media	City Committees, interviews
Number of actors involved	2000 citizens and 100 local associations	60 local associations		5000 interviewed citizens
Issues dealt with	Transports, air, employment, houses, culture and spare time, open spaces, educational activities, public health, energy, rubbish	Energy, building, rubbish, traffic and transports, nature and open spaces, health, education and information	Lifestyle, sustainable cities, socio-economic and environmental systems, environmental co-operation among towns	Water and air pollution
Documents drawn up	The process is still in progress	Environmental Action Plan Set of indicators	Action Plan for sustainability	Environmental Action Plan
Established objectives and targets	The process is still in progress	Mainly environmental objectives	Very precise objectives and targets of a qualitative and quantitative nature	Environmental objectives of qualitative and quantitative nature
Process aims	Combine development and environmental protection in a long term perspective	Change in population lifestyle, through a participative strategy, improvement of environmental services by the municipality	Change in population lifestyle through a participative strategy, improvement of environmental services by the municipality	Dissemination of knowledge on environmental issues, improvement of environmental services by the municipality

fundamental element for the initiative's success. The practical experiences, on the other hand, differ from each other, especially in the issues tackled. Indeed, some cities focus the process exclusively on the environmental problems whilst others consider the economic and social variables as well, thereby embodying more correctly, perhaps, the concept of sustainability. Moreover, further differences are seen both in the type of documents drawn up and in the nature of the objectives established. These objectives can vary in their degree of precision, and can be short- or long-term, qualitative or quantitative. The action plan can, therefore, assume different forms. There are very detailed documents which indicate the objectives' deadlines, which define precise targets and identify the actors responsible for implementing the strategies. In other cases, on the other hand, the action plan is limited to giving general indications, by establishing objectives of a qualitative nature. Finally, it is important to emphasise that the aims stipulated for the Local Agenda 21 process influence both the choice of instruments to be used and the type of document which is obtained.

2.3. A possible classification for Local Agenda 21s

The study of the international experience of Local Agenda 21 permits the establishment of a possible Local Agenda 21 classification. This classification makes it possible to distinguish the various initiatives in relation to the real or presumed, the direct or indirect aims which the Local Agenda 21 process is able to pursue.

As has already been said, in many cases, what causes the local authority to set up an Agenda 21 process is almost exclusively pressure from environmental groups who become included in the initiative in order to avoid or reduce probable conflicts. It is clear, therefore, that in such a context, there is the danger that the policies embarked upon have a purely symbolic significance and the Action Plan obtained is essentially worthless. In these cases, it is difficult to succeed in fitting the exercise into any classification. Nevertheless, the origin of the process must not be confused with its content. Indeed, nothing prevents the Municipal Administration from modifying its own objectives in the course of the process.

In general terms, a possible classification identifies the following forms or pure models of Local Agenda 21:

(1) Local Agenda 21 'of knowledge': the principal aims of this type are the creation of a cognitive base and its dissemination amongst the population. Such objectives are generally pursued by means of drawing up of reports on the state of the environment and by establishing sustainability indicators with varying degrees of participation. This is a form of Agenda 21 which is very common. Furthermore, in any case, the creation and dissemination of knowledge on the environment are results found in almost every process.

(2) Local Agenda 21 'of involvement': is certainly the most wide-spread form abroad, particularly in the United Kingdom and in North America. Its objective is to stimulate, by adopting a participative approach and using communication instruments, a change (sometimes more and sometimes less gradual) in the population's behaviour and life-style through developing environmental awareness in families and businesses. This type requires the adoption of a participative approach and the use of communication instruments which can involve the highest number of local citizens, organisations and groups.

(3) Local Agenda 21 'of improvement in services': in this case the local authority uses Agenda 21 as an instrument for reviewing its own plans and programmes, for re-evaluating its own functions and for attempting, as a result, to improve, within an environmental framework, the provision of services for the community. The document obtained, therefore, mainly addresses the municipal administration, the municipality's department of the environment which regulates the behaviour of private individuals, or municipal businesses which provide services such as waste disposal, water supply and public transport.

(4) Local Agenda 21 'par excellence': in this type of Agenda 21 the process's aim is to be able to combine environment and development within a long-term perspective. This is an Agenda 21 which, although it does not give up the possibility of being put into effect immediately, proves to be heavily oriented towards economic and social development and towards its connections with a sustainability approach. It is not easy to find this type of Local Agenda 21 in practice even if several cases seem to show signs of approaching it in some respects. We refer in particular to the cases of Hamilton-Wentworth and Seattle, where the definition of a shared vision of sustainability in the long-term is considered an indispensable pre-condition and the starting point for the entire process.

2.4. Lessons from Agenda 21 experiences

The survey conducted makes it possible to highlight several points, both positive and negative.

The Local Agenda 21s constitute a fairly composite phenomenon in which a series of disciplinary traditions, methodologies, concerns and disparate objectives come together. Although the phenomenon is already some years old, and despite the fact that various attempts to this end have been made the definition of a single model does not yet seem to have emerged in international practice. Even in the sample which we have examined, this diversity is clear, and not just in the different nature of the cities considered (obviously London is not Troyan in Bulgaria.) From this point of view, an initial lesson which concerns us directly is that Venice's unique nature well justifies a different approach without needing to abandon the name of Agenda 21.

This does not mean that the experiences have nothing in common. The most visibly unifying element is the absolute centrality of the environmental concern, which in some cases means its exclusivity. More simply, whilst in various cases there is little or no talk of development, and of economic development in particular, not a single case exists, on the other hand, in which the environmental issue does not occupy a totally predominant position. The adoption of Agenda 21, therefore, entails constraints which, as will be seen, can also be considered as an opportunity: i.e., one of placing the question of environmental protection at the centre of the definition of Venice's future. It must be stressed that a similar option would change the local and national debate about Venice's future quite radically.

For this reason, it must also be noted how the notion of the environment which prevails in the Local Agenda 21 experiences is often an ambiguous one, even if the notion of sustainability is almost always central. The prevailing meaning of sustainability seems to be one which refers to the dimension of intergenerational equity; more simply, what is assumed is the requirement to pass the natural heritage on to future generations as it was inherited by previous generations. The ambiguity, often desired, lies in the relationship between the local dimension and the global one. Indeed, on the one hand, the emphasis on taking action which constitutes a fundamental feature of the exercise gives rise to a focusing on the local environment and on how to improve it here and now. On the other hand however, the very origin of the experience, its beginning, that is, in connection with concern about global change, compels, in many cases, an emphasis on the local contribution to protecting the global environment; for example, on reversing the trends towards climate change. In other words, the effort made is often that of constructing positive externalities subjectively and willingly, thus going beyond the free-riding practices.

A corollary of this dimension of volition is the fact that one of the objectives which is most clearly and most often found in Local Agenda 21s is the change in the population's lifestyles, which, in turn is brought about or, at least encouraged, by an educational/participative process which tends to directly involve a consistent number of citizens. The more obvious examples of this are the changes in behaviour as regards the production and collection of waste, energy consumption and individual mobility.

Of greater importance, however, are two almost omnipresent elements. These are the involvement of social organisations, and not only environmental ones, and the construction of a scientific base, exemplified by the system of environmental indicators. These two elements are deemed to be instrumental in a change in policies which is obtained by means of the change in society's demand for environmental protection. The principal planning approach is that of a political/administrative actor (almost always the municipality) which attempts to set up a vast network of actors as a means of involving the population in order to change its perceptions and demands, with the aim of reorienting policies and social behaviour for improving the environment in the direction of global sustainability. Therefore, it is a matter of a strategy created by local government to produce a global impact.

If all this is true, it follows that the international Local Agenda 21 experiences are only able to create an adequate reference point up to a certain point in building the Venetian process. The largest difference is formed by the need in Venice to confront the problem of the robustness of the model for economic and social development in the long-term, given the intrinsic contradictions of the industrial model, on the one hand, and the tourism model on the other.

Furthermore, the global/local relationship in Venice seems to be objectively different: Venice is, in itself, a global problem and is threatened by global changes. Therefore, rather than asking what the Venetians can do to improve the planet's environment, it appears necessary to ask what can and must be done (in Venice, but not only in Venice) in order to pass on to future generations a Venice which is probably different from today's and yesterday's Venice but one which is, possibly, no less able to reproduce its order (social, economic and environmental).

Even with these reservations, the survey carried out is anything but useless. In fact, the lessons which can be drawn from the international Agenda 21 experiences and which constitute numerous elements for planning are the need in each case to link development and environmental protection, to adopt an intergenerational notion of environmental sustainability, and to strengthen the local/global connection. These elements, which seem to complicate the governmental processes which already seem quite complex, actually make it possible to change the nature of the game and make it more manageable after all.

3. Strategic planning and 'coastal zone management'

The characteristics of the Venetian case lead to a change in perspective away from the mere imitation of Agenda 21 processes which predominate abroad. It is on the basis of these considerations that we have redefined the problem and have searched for other models, turning our attention to strategic planning experiences. The notion of strategic planning adopted here, which is, after all, in line with more mature consideration, is composed of various elements:

- first, as is natural, of a long-term temporal dimension;
- second, of an emphasis on the need to build consensus between the actors;
- third, of a model based on the need to achieve the planning objectives *in an indirect way* by changing the perceptions and behaviour of society's actors, and not only by using authoritative instruments.

The research procedure which we have followed is to investigate the experiences of strategic planning which are available and accessible at the international level in order to see whether and to what degree these can help us to identify the elements necessary for planning the Venetian process. The research questions were related to the ways in which local planning policies are articulated and, more broadly, to an area's territorial management should these policies intend to combine considerations tied to economic development with others bound to environmental protection. In this way, several tentative conclusions can be drawn which provide numerous ideas for the debate on the Venetian area. Obviously, it is not so much a matter of transferring from different contexts methods of action which are tied to other systems of actors and to different ways of structuring the problems, as of, rather, opening up some questions and of exploring them by using a multiplicity of perspectives, supplemented, in this case, by the survey conducted.

The first conclusion which can be reached is the fact that in ordinary strategic planning at an urban or metropolitan level, whether or not they might use a SWOT model (strengths-weaknesses-opportunities-threats) or any other model, the environmental question does not seem to be included in the elements which are deemed able to improve urban competitiveness. At most, the environment is considered to be one of the amenities which contribute to determining the quality of life, an element considered so important as to make managers of large firms decide on the location of their production or administrative activities.

The focus was turned to planning procedures which could take the problems of the environment and of economic development into account right from the start, if for no other reason than to examine the actual possibility of articulating the two issues satisfactorily, combining, at the same time, long-term vision and emphasis on taking action, decision-making effectiveness and consensus between the interested parties. The examination made has shown that it is certainly not easy to obtain similar objectives but that, at least as far as the Dutch ROM (see Box 1) and the air quality plan in Southern California are concerned, that it has actually taken place, despite the existence of serious conflicts and of fragmented jurisdiction.

Indeed, the most important lesson which emerges from this investigation concerns precisely the greater effectiveness which government efforts have if they do not attempt to reduce the decision-making complexity artificially, but, instead, explicitly accept making different functions of use compatible, the co-existence of interests in conflict and the plurality of actors as an issue and a problem to be dealt with alongside the issue of the objectively more suitable solutions. Indeed, at the stage when all of the relevant problems must be brought into the process, at the same time, given the close link which is produced between them, it becomes necessary to bring all of the actors who represent the different demands into the process as well. In general, this forms an initial step towards starting up a dialogue which could then become a real negotiation process.

A second important point concerns the actors' involvement both in the decision-making process and in the implementation phase. In some cases, this issue is widely discussed, in other cases is is not explicit, but it does not prove to be less important for this reason. In terms of a negotiation, within which the public actor necessarily takes on a role which is radically different from their usual one, the involvement of all the relevant actors is a highly delicate and crucial point of the whole process. In fact, the central public actor must, at least partly, abandon its own traditional role of directing a planning process within which the private actors are usually called upon for carrying it out. To a large extent, the success of the policy adopted depends on the correct structuring of this change, whether it is a matter of involvement which originates from above, or whether one of construction from below of decision-making and operational networks.

In the final analysis, we can consider the relationship which becomes established between the public actor, in the guise of initiator and director of the negotiation processes on the one hand, and the different stakeholders on the other, to be an extremely important one. The most important and most extensive body of literature and experience on these subjects comes from coastal zone planning (see Table 2). This is both because it is a subject with the longest experience and which, in many cases, precedes the growth in environmental concerns and also, and above all, because the co-existence of sea and land, and of the potentially conflictual functions relating to their use, has actually made any reductionist approach impossible and has forced the actors to take the whole decision-making complexity into account.

In the European and, even more so, North-American experiences of coastal zone management, the need to bring into the decision-making process all of the actors relevant to that process, within a generally coherent model, usually becomes a factor for success. This is because, in this way, no topics can be ignored, but, rather, all relevant topics become the subject of negotiation, whether they be those relating to economic development or those concerning enhancing the environment's value.

The most significant methodological outcome which emerges from the analysis made concerns the fact that the most rational approaches (cf. the OECD model) have essentially failed or, at any rate, have encountered great difficulties due to the impossibility of establishing long-term forecasts and to the cumbersomeness of instruments for institutional co-ordination. When, instead, the road of voluntary agreements has been chosen – agreements which are often submitted to open scrutiny by the population and social parties and are always supported by information, monitoring and evaluation systems for the different physical and socio-economic components – exemplary mechanisms are triggered for co-operation on and internalisation of the mutual interests and objectives, on the part of the different actors, which have allowed significant advances to be made towards the act of promoting development and environmental protection.

In other words, the analysis of cases of strategic planning and of coastal zone management gives us a series of precise suggestions. The most important of these is that a long-term strategic planning exercise must contain all the topics considered important by all the actors whose participation is necessary for the initiative's success. We shall return below to other aspects, but for the moment it is enough to draw the conclusion, in many respects intuitive but all too often

Table 2 Coastal zone management experiences

Involved areas *Process Characteristics*	*Frazer River*	*Barataria-Terrebonne*	*Great Britain Southern Coast*	*Thames Gateway*	*Great Lakes*
Start date	1977	1990	Various	1994	1978
Process phases	FRES I, (1977/78) study phase FRES II (1978/82) decision-making structure planning phase FRES III (1983/84) project phase	Signature of a starting *Conference Agreement,* state of the art analysis, drawing up of an integrated management plan, starting of single action plans	In one case, the process is started with the opening of a negotiating table between the *County Council* and British Petroleum; in the second case the start coincides with the unification of management competence under the Ministry for Agriculture, Forests and Fishing	Drawing up of a Strategic Plan	Different policies adapted to reports issued every two years by the *International Joint Commission* which address the action programmes
Promoters of the initiatives	British Columbia *Province* and Canadian Federal Government	Louisiana Country and *Environmental Protection Agency*	Ministry for Agriculture, Forests and Fishing, *County Council* and *District Council*	Ministry for the Local Government and Planning	United States and Canadian Federal Governments
Process issues	Soil use, transport and harbours, water quality, state of natural habitats, recreational activities	Water quality, natural resources, estuarine uses, in relation to different pollutants	Oil and natural gas extraction plant location, protection from erosion, improvement in bathing water	Economic revitalisation, physical adjustment and environmental enhancement	Reduction in pollutants, physical and morpho-logical protection of water bodies, food chain restoration
General aims	Conciliation of economic development based on heavy industry, wood industry and harbour activities, and protection of a complex coastal environment	Restoration and start of sustainable management of estuarine resources	*Trade-off* management of economic development and environment	To boost the global revitalisation of the area, starting from the importance of the river from both a productive and natural point of view	To restore and maintain chemical, physical and biological integrity of waters belonging to the Great Lakes Basin Ecosystem
Decision-making structure	*Management Committee* (Ministries for the Federal and Provincial Environment, Harbour Activities) which co-ordinates the activities	*Policy Committee* and *Management Committee* elected by the Government, assisted by *Scientific-Technical Committee, Citizens Advisory Committee* and *Local Governments Committee*	The decision-making structure is the ordinary one of the management of the territory	Thames Gateway Forum chaired by the Ministry for Local Government, with local authorities and government agenices; on the sub-areas level, *fora* with actor involved	*International Joint Commission* with some research organisms build up guidelines informing on local administration and agencies action

forgotten, that any system for governing complex situations must, in some way, reproduce the complexity of the problematic situation and must not seek, instead, improbable simplifications through the imposition of superimposed points of view. It is precisely because of the dramatic nature of the problems and because of the severity of the current and potential conflicts that a successful strategy must choose, both decisively and patiently, the road of agreement and confrontation.

4. Decision-making complexity in policies for Venice

4.1. The need for an empirical analysis

If this is true, it follows that the planning of a decision-making system which is adequate for simultaneously dealing with development problems and with economic revitalisation in Venice, as well as with environmental protection, in its concurrently global and local dimension, can only begin by attempting to recreate the degree of decision-making complexity which is specific to the Venetian environment.

It is worth dwelling on this point for a moment. As is well-known, Venice has a series of considerable peculiarities in terms of institutions as well. The special legislation which, at least since the flood of 1966, has accumulated in the regional capital on the lagoon has, in fact, created an institutional network based on three pillars:

(1) The existence of an Interministerial Committee or *Comitatone*, chaired by the Prime Minister and in which, as well as the central administrative organs, the regional and local authorities are also represented. This holds the power of approval for all intervention financed with the subsequent special laws;
(2) The existence of a single concessionary, the Consorzio Venezia Nuova, which is a consortium of private firms whose responsibility is the execution of works of state responsibility where the protection of the city and the lagoon are concerned;
(3) The Protection Commission, or *Commissione di Salvaguardia*, an administrative organ, appointed by a multiplicity of central and local bodies, which is responsible for the approval of all the town-planning instruments and of construction permits, except for those which are the *Comitatone*'s responsibility.

It is common opinion that the combination of these institutional structures has not stood the test, if for no other reason than because the co-existence of the extraordinary activity assigned to these institutional structures and their normal responsibilities has created more problems than it has resolved. Nevertheless, from this diagnosis, which is very wide-spread amongst the observers, not a single general prognosis has been derived.

With respect to the demands mentioned in the previous section and which concern, among other things, the need to deal with questions of a different nature simultaneously, the traditional philosophy which has inspired the institutional structure of the government of Venice appears very far away. It appears removed, above all, from any examination of the reality of what it means to govern Venice, and it is, instead, broadly based on a model which is essentially valid in every place and in every situation.

Compared to this approach, the position which we intend to adopt here is radically different. In light of the challenges of extreme complexity which are set by the establishment and, above all, by the implementation of a environmentally sustainable model for social and economic development, in fact, it seems necessary to base any project for restructuring the decision-making system on a comprehensive identification of the parties involved, of their mutual interrelations, of the interests and values which manage to reach the decision-making arena, and of the available resources.

What has just been said is not a usual approach in the literature on territorial governance even in countries which have a better and longer tradition in the matter of collaboration between different actors than we have. Furthermore, if it were possible to isolate the precise elements of what it means to govern Venice, perhaps the procedure which we are proposing would not even be necessary.

Unfortunately, this is not the case. The question of social and economic development in a context of environmental sustainability is such a complex problem that it appears completely over-ambitious to wish to isolate it. To develop a Local Agenda 21 in Venice means neither more nor less than defining the future expected and providing for the necessary projects as well as the means of starting them. This is exactly the definition of government.

In short, the point of departure for an action plan for sustainable development undoubtedly consists in the creation of a vision and of a procedure. However, its efficacy also, and particularly, depends on its capacity for determining a series of projects which are varyingly linked with each other and are put into effect by a multiplicity of public and private actors.

Our reasoning is that from a reconstruction of the network of actors which already participates in urban transformation in Venice today, there should emerge useful indications on the features of a possible process of strategic planning orientated towards development and environmental protection.

In more abstract terms, the fundamental assumption is that it might be possible and reasonable to outline and establish the system for governing a complex situation by analysing the set of the individual decision-making processes which take place in the situation itself. In other words again, the solution of the 'macro' problem, the government of Venice, depends on the features of the 'micro' processes, or the individual projects, which the macro problem ought to bring into being. Just as the reconstruction of the Venetian economy is based on the idea that the individual Venetian production activities constitute the basic components of the economic system, so our reconstruction will also be based on the individual projects and actors which participate in them in order to design the system for governing Venice.

4.2. Projects for Venice

The point of departure for this reasoning is, therefore, a list of transformation processes for which the actors are known. The projects are considered to be the minimum observable unit of the whole policy-making process and the hypothesis is that, if there are enough of these projects, the sum of the actors constitutes a significant representation of the protagonists of the governmental process in Venice.

Obviously, the ideal for carrying out such a research programme would have been to start from zero and to reconstruct the projects for transforming the city by using a method which is homogenous to and consistent with the research objectives. This, however, would have involved considerable cost and, above all, would have delayed the outcome of the work.

Therefore, we have used as a basis the high-quality work carried out in the Department of Economic and Social Analysis of the Territory (DAEST), financed by the ex MURST 60% funds, and directed by Professor Pierluigi Crosta. The material collected within the scope of the research project 'Policies-Venice' concerns precisely the projects which have been carried out or are planned in Venice in recent years. For each of these projects, the researchers of the DAEST analysed the daily press and periodicals from 1995 onwards. In many cases, this was completed with the collection of official documents and with some general interviews with the relevant actors, so as to create summary records based on a common model. In some cases, this information was expanded on for the purpose of our work, but, as a whole, it was sufficient for the analysis which we intended to carry out.

By applying a series of criteria, which we will see immediately, a list of 30 projects was selected (see Appendix 1), which includes a series of decision-making processes taking place around the middle of the 1990s in Venice. Included in this definition of Venice is the historical centre, the islands and the lagoon, as well as the lagoon shore-line which goes from Porto Marghera to the airport of Marco Polo, passing through San Giuliano. As a result, projects for urban transformation in Mestre remain excluded.

An initial selection criterion was formed by the availability of information for a period of about one year, in order to minimise the risks of having information which was incomplete as regards the set of actors involved. However, if the individual projects were included within a more complex project, we preferred to keep to the latter rather than to the individual constituent elements (see, for example, the Giudecca project which incorporates a series of works in the sector of residence and of economic development).

Furthermore, particular attention was paid to ensuring that the various projects concerned the principal problems in Venice as much as possible, that is, residence, socio-economic development, environmental protection, increasing the historical-artistic heritage's value, etc. In purely numerical terms, initiatives for economic development and environmental protection tend to predominate but culture, transport and residence are also sufficiently represented.

On the other hand, whether the projects were similar in terms of the stage in the decision-making process was not one of the selection criteria. We thus find cases still at the idea and proposal stage, others in which a decision-making phase has already been reached, and others again which are being implemented, or actually in the concluding stages. One of the hypotheses to be tested with a similar variance of the sample is whether the number of actors involved and, therefore, as we shall see, the complexity, tended to grow as the concluding stage was approached.

The result is a broad range of initiatives which differ greatly from one another in terms of their importance (going from works for defence against high water, probably one of the greatest hydraulic engineering projects ever conceived of, to the regulation of clam fishing in the lagoon) or in terms of the degree of conflict between the actors, and which are found within various sectoral policies and which are at very different stages of development (from the City of Music, little more than a planning idea, to the Science and Technology Park which is being fully realised). We believe that such a sample can legitimately be considered representative of the governmental processes in Venice.

4.3. The Complexity of Governing Venice

The next stage of the analysis regarded the identification of the actors who are protagonists of the Venetian processes. As a result of this, a list was produced which includes about a hundred actors who, for greater clarity, have been classified in a double entry table which takes both the territorial level of reference and the type of prevailing rationale into account. The first part of the analysis is the dimension of the interests and problems involved, while the second is the nature of the issues and the problems which are put forward. In short, in every cell of the table there is a kind of specific actor all of whom can have significant conflicts and contradictions between themselves, but which discuss, at least potentially, at the same level.

As always when there are complicated problems of classification, the allocation of an element to one cell rather than to another is rather arbitrary. Nevertheless, even allowing for a certain degree of inaccuracy, several elements are clearly evident. The main feature to which we shall have to return, is the fact that the actors seem to be distributed along a diagonal which goes from the corner formed by the international institutions to the corner made up by social interests of a local dimension. This finding is reinforced further by the fact that the high presence of economic interests at a national level seems to be substantially misleading: (as many as five of the actors listed appear in only one process, the one for the centre of marine technology, and not necessarily with a significant role). Another conclusion can be derived from this: at the national level, a political-institutional dimension tends to prevail, whilst the technical-scientific actors become denser at the regional-provincial level and at the city level and, as we shall further emphasise, the social interests prove prevalent at the local level.

The classification permits us to take a step forward in the analysis since it suggests reasoning in the following way. Even leaving the number of actors involved out of consideration, the level of complexity of a decision-making process is represented by the number of different types of actors involved, as defined above by the relationship between the dimension of the interests and type of rationale behind the action. It can be inferred that a process which sees the presence of international economic actors, national political-administrative institutions, local social groups and also national experts is richer and more differentiated than one which is carried out completely at the same level, the city level, for example, and between actors which are similar to each other (the parties represented in the municipal council.)

Thus, we have envisaged a complexity indicator which is obtained from the product of the number of different levels of territorial reference for the actors involved in a single project and the number of types of actors beginning from the nature of the attributed interests. It is clear that this indicator varies between 1 (if all the actors belong to the same category and operate at the same territorial level) and 20 if there is at least one actor for each territorial level and for each type.

The result of this analysis in relation to the sample of projects considered is shown in Figure 1. As can be seen, the projects surveyed are placed on a scale from 2 to 16. In the first case, two territorial levels of the same kind of actors are represented whilst at the completely opposite end of the scale, all types of actors which are distributed across four territorial levels are present.

The analysis of the indicator confirms, once again, how the research has probably underestimated the decision-making complexity, not so much at the top end as, rather, at the bottom. The simplest processes are often also those at a low level of visibility and/or of development in terms of planning. Typically, they are projects still in the initial stages which sometimes have not even gone beyond the idea stage.

The contrary is not true. In fact, among the most complex processes we find cases in which the project's actual start date is uncertain and, in any case, remote. On the whole, they are projects for environmental restoration and/or for re-use of large lagoon shore-line areas; i.e., the kind of decisions which Venice would probably need and which should form some of the cornerstones of a process of strategic planning and of Local Agenda 21.

Nevertheless, the measure of decision-making complexity makes it possible to clarify an important aspect. Indeed, as can be seen, the mean value of the indicator lies between 9 and 10. In other words, the average project sees the

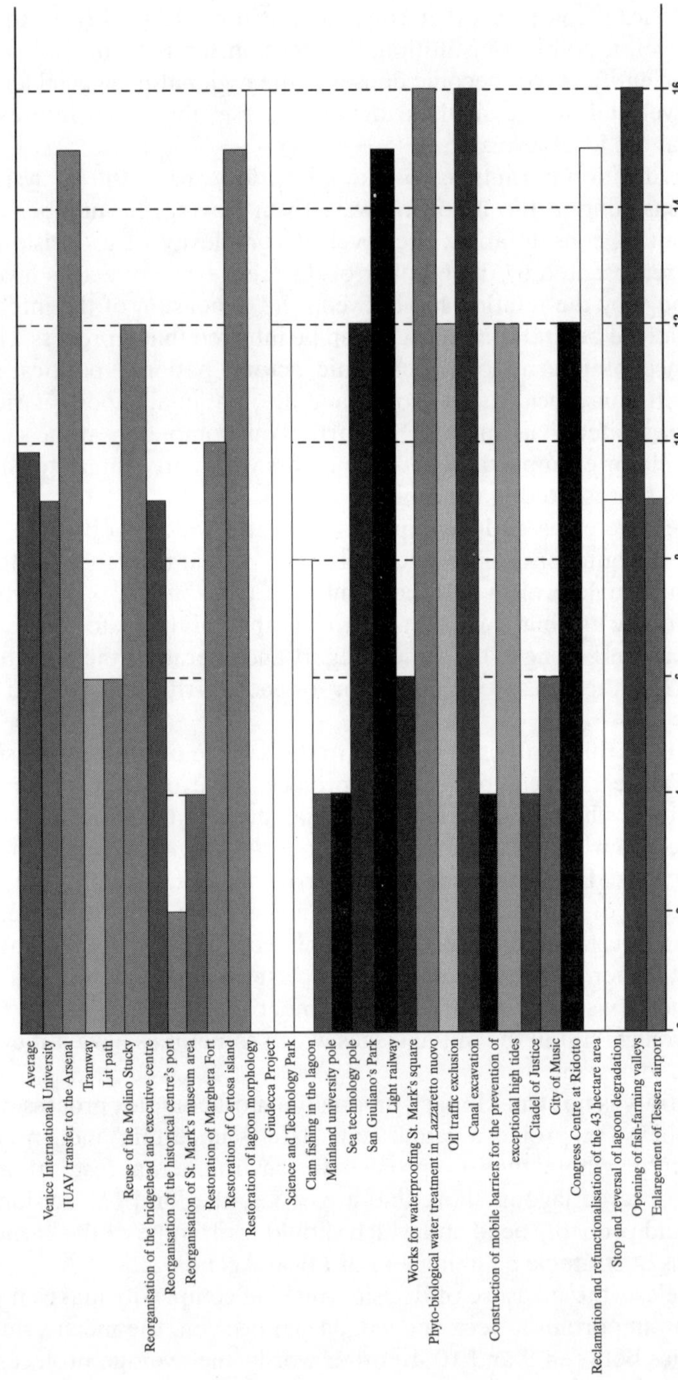

Figure 1 Complexity of surveyed projects.

presence of three types of actors at three different territorial levels or of two types of actors at no less than 5 different territorial levels. In short, any governmental endeavour which does not tackle the issue of the intergovernmental dimension risks omitting an aspect which is central to the problem. The governance of Venice cannot be exercised in Venice alone, but necessarily entails a link with a dimension higher than that of the city.

4.4. Does a 'government' exist in Venice?

The next analytical step deals with the need to simplify the picture and to identify the central actors in the process of urban transformation. This exercise had two key sides: on the one hand, the identification of actors who participate more frequently in the projects surveyed and, on the other, the reconstruction of links which are established between them.

Both analyses, however, require a simplification of the field. If, in fact, the level of detail contained in Table 3 is retained – a table in which all the individual divisions of the institutional structures and all the individual interest groups are correctly indicated as actors – a picture is produced which is essentially illegible. In fact, more than two-thirds of the actors are present in only one project and, out of the roughly 5250 possible pairings of the same actors, only a little more than 1% actually occur.

Therefore, a series of adjustments has been made to the list of actors. First, in the case of the regional and local institutions, these have been taken as single actors. Second, two categories of actors have been created. These are the technical agencies (which group together the municipal companies and, more generally, the support structures outside the local administration) and the environmentalists (which group in a single category all the ecological organisations and the Green Party at a local level.) However, both analytical adjustments leave the level of complexity as it has been defined above unchanged, once the unifications have been made between actors which are in the same cell in Table 3. Even with these simplifications, the total picture is still a long way from being easy to read. Nevertheless, the total number of actors drops to about 60% of the original sample (i.e., to 44) and 3.8% of the roughly 950 possible pairings actually occur. An initial conclusion, therefore, regards the fragmented and dispersed nature of the network of actors which governs Venice. To simplify the situation further, we have concentrated on those actors who participate in the decision-making processes more often, fixing as a threshold their presence in at least 10%; that is, that they participated in at least 4 projects.

The result is summarised graphically in Figure 2 and shows a significant drop in the number of actors. The 18 who remain and who are globally present 157 times, therefore, have been grouped together into four quadrants by differentiating, on the one hand, the institutions and technical bureaucracies from the economic and social interests and, on the other, by differentiating the

Table 3 The actors involved in the projects surveyed

	Institutional–Political bureaucracy	*Experts–Technical bureaucracy*	*Economic Actors*	*Social Actors*
International	European Union: DG XII DG XVI		IBM Semea Ch. Forte Group Eni	Greenpeace International Committee for the Conservation and Protection of the *Antico Arsenale* (Arsenal)
National	CIPE Inter-ministerial Committee ex. L.798/84 MURST Ministry of Cultural and Environmental Heritage Ministry of Transport Ministry for Public Works Ministry of Industry Ministry of Finance: State Property, State Monopolies Ministry of Justice Ministry of the Environment Trieste Harbour Ravenna Harbour	Ministry for Public Works: CER Superior Council for Public Works Ministry of Defence: The Navy FFSS (State railways)	Consorzio Venezia Nuova ICIP-Cameli Alumix Group Edison (Montedison) Fincantieri (with Bonifica s.r.l.) Tecnomare Efimimpianti Alenia Assicurazioni Generali Istituto Bancario S. Paolo di Torino Acqua Marcia S.p.A. Ridotto s.r.l. (Benetton Group S.p.A. Telecom	Italia Nostra WWF
Regional/ provincial	Veneto Region: Regional Council, Industry and Energy Department, Dept. of Agriculture, Hunting and Fishing Province of Venice: Regional Council, General Secretariat-Presidency Harbour Office Archbishop Curia	Ministry of Defence, North-eastern region Division Ministry for Public Works: *Magistrato alle Acque* (Water Magistracy), Magistracy Technical Committee Ministry of Cultural and Environmental Goods: Government Office for Fine Arts, Government Office for Archaeological Goods Veneto Region: Transport Office Province of Venice: Ecology, Hunting and Fishing Office, Offices for Defence and protection of the the territory, Surveillance, Civil Protection, Technical Office, Contract Office, Heritage Office Protection Commission USLL • IACP • CNR Veneto Innovazione SpA	SAVE (Airport company) Industrial associations CARIVE Foundation *Consorzio per lo Sviluppo dell'Università di Venezia* (Consortium for the development of the University of Venice)	Forum for the Venice lagoon Italia Nostra WWF

Continued …

Table 3 Continued

	Institutional–Political bureaucracy	*Experts–Technical bureaucracy*	*Economic Actors*	*Social Actors*
Municipal	Venice Municipality: Municipal Board, Mayor, Municipal Council, Officer of Heritage, Officer of Culture, Officers of Public Works, Special Law, Green areas and Parks, Accounting and Budgets	Venice Municipality Special Law Office Town-planning Office IUAV University of Venice (Ca' Foscari) ASPIV • EdilVenezia • ACTV • AMAV Immobilaire Veneziana Venezia Servizi S.p.A. Promomarghera	SVIT	Committee for the Protection of Porto Marghera Environmentalists Greens Partito d. Rifondazione Comunista (Communist Party)
Local	District Councils Mira Municipality	*Centro di Recerche e Restauri di Palazzo Cappello* (Cappello Palace Centre for research and restoration)	Caffè Florian and Caffè Quadri Consortium for Venetian small shipyards Consortium for Fishermen Lagoon 'Valley' Fish Farmers Figheri lagoon 'valley' governors	Nautical and Rowing Association Administrative Lawyers Association Archeo club/Ekos club Venetian Cultural Association *Tera & Aqua* (Land & Water) Association Certosa Committee

supra-local actors from the local ones. As is shown graphically, the most crowded cell is that of the supra-local institutions with 10 actors (55% of the total) which prove present 74 times (47% of the total). The cell of local institutions follows with 5 actors (27%) present 59 times (38%). Particularly empty, on the other hand, is the column of interests where at the supra-local level there are only two actors (the *Consorzio Venezia Nuova* and ENI), only present 11 times and, at the local level, only the category of environmentalists with 13 presences.

If we go on to analyse the individual positions, the fact which stands out is that the municipality of Venice is present, with one or more of its internal actors, in no less than 23 projects out of 30 (almost 77%). Placed at a considerable distance, but still always with a conspicuous presence, there are the environmentalist movement (present in 43% of cases) and the *Magistrato alle Acque* (present in 40% of cases).

Some of the most interesting data includes the presence of the municipal technical agencies in almost one-third of cases, which corresponds, in our opinion, to a precise strategy on the part of the municipality to exploit these kinds of actors in the planning and implementation of innovative projects. Equally significant, with 13 presences, is the role played by the Venetian universities. This is certainly due in part to the fact that they were amongst the beneficiaries of financing in the last special law (law 139/1992), but indicates, perhaps, more generally, their active role in city life, which is linked to the fact that the expansion in higher education and research activities is one of the few trends compatible with the constraints represented by the Venetian situation. However, the most significant fact which emerges remains the extreme under-representation of economic and social interests. Above all, out of the former, the absence from the debate of the trade associations of institutions such as the Chamber of Commerce, and of the most important firms stands out.

Out of the social interests, the absence of trade-unions must be pointed out and also that of political parties which do not seem to take a clear and, in some way, official position on most of the issues. If this is partly understandable for the parties in power, which can perhaps communicate internally with the municipal council (but we would not bet on a similar hypothesis), what appears more difficult to explain is the absence of opposition parties and also representatives of other social interests. Among other things, it is necessary to emphasise how the sources of information used by researchers, that is, in general, the daily press, usually tend to over-represent precisely this type of actor.

4.5. The Map of Actors

First, however, it is necessary to make a further analytical step, the last in the whole sequence, and to tackle the problem of links between the actors. To this end, we have simplified the picture further by limiting the analysis to those actors who were present in conjunction with other actors at least four times. In doing this, the number of central actors dropped from 18 to 14. The Port Authority, the Province of Venice, the district councils and ENI, who were present four or five times in the projects, but, obviously, not when linked together with the same actors actually disappeared off the map. Essentially, this means that this type of actor tends to participate in projects which are, on average, less complex and probably more specialised. Therefore, such actors tend not to have strong links with the other principal actors in the government of Venice.

The results are summarised in Figure 3. Here, as well as the number of times present, the links between the actors are shown as well as are the intensity of such links. As it is easy to discern, this is the most cohesive network out of those considered hitherto since, out of the 182 possible pairings, 23 are registered which is equal to 12.5%. Moreover, it must be noted that this is a largely underestimated figure, since only four or more links are recorded.

The most surprising presence, both for its frequency and the intensity of its relations with the municipality, is that of the European Union which participates in almost a third of the processes. This participation is invariably in the form of financing managed by Directorate General 16 which deals with regional policy (it must not be forgotten that Venice is included in Objective 2 of regional policy which refers to areas undergoing a process of de-industrialisation). Moreover, the existence of special legislation for Venice undoubtedly encourages the mobilisation of state financing. However, the question of financing alone does not appear to be sufficient to explain EU presence, given that in the lagoon city there is certainly no lack of money for investments. If anything, it is the capacity for deciding on the investments and putting them into effect which is lacking. Therefore, the most convincing interpretation appears to be, firstly, that the involvement of the EU Commission increases the project's acceptance, in the sense that, in the absence of a shared vision for development, a project's access to international financing increases its legitimacy in the eyes of the local actors. Second, however, one has the impression that another element which makes the EU involvement valuable is the greater rigidity of its formal procedures; i.e., the existence of predefined deadlines for applications and for the subsequent decision-making and implementation. This greater stringency is used by the project promoters to remove the opposition's chances to postpone and, more generally, it automatically selects the initiatives with greater support and with fewer conflicts. In this way, the strong link between the municipality and the EU forms a true and proper strategy, on the part of the Mayor and the Council, which tends to maximise participation in European projects also as a means of getting round the decision-making difficulties which are inherent in a model like the one described.

A further observation which is worth making concerns the collective bodies and especially the Protection Commission whose involvement, notable for its frequency, does not seem to be able to reduce the overcrowding of the network. In other words, this organisation has the feature of being another actor which supplements those already in existence but which does not replace them. This is clearly visible if a link is seen to exist between this commission and the government offices which are also represented in the same Protection Commission. This means that the government offices, although the observation can clearly be extended to other actors, do not believe that their participation in that body exhausts their role fully and they tend, therefore, to act on their own as well.

Apart from that, there is nothing more to add, except for noting how the centrality of the municipality is confirmed by the fact that it has strong links with no less than nine of the 13 actors present. The four absences are the *Comitatone* (in which the municipality is one of the main members), the *Consorzio Venezia Nuova* (where there is a clear division of labour), the technical bodies of the Ministry for Public Works (a completely understandable absence and, in any case, one of marginal importance, especially as there is a considerably strong link with the *Magistrato alle Acque*) and, finally, the University of Venice (which probably depends on the choice of cases). There is enough to conclude that the municipal administration is indisputably the pivot of the entire process.

4.6. The complexity of governing Venice: an initial conclusion

In our opinion, the study carried out demonstrates the real possibility of analysing the characteristics of the governmental processes in a way which is not exclusively based on impressions and qualitative, but which tries, instead, to identify measurements of and ways of representing the complexity which can be reproduced and compared both temporally and spatially. If properly treated, the breadth of and differentiation in the sample of cases considered gives an entirely realistic picture of what it means to govern Venice as well as of the problems involved in such an exercise. Further investigations which ascertained the different sets of actors involved on the basis of other sources, perhaps even managing to specify their role in the different decision-making processes, would certainly be useful. However, the work which has already been carried out is intended to indicate a method for saving the debate from getting stuck in abstract institutionalism as well as a way of summarising and representing a situation which is not destined to change in the short- or medium-term unless improbable palingenesis is envisaged.

Furthermore, it is necessary to stress, as has been repeated many times, that the analysis made in the preceding pages is constantly oversimplified: fewer projects were considered than the total number of projects and the set of actors for each one of them could probably be extended by a more in-depth investigation.

Firstly, the most positive conclusion which can be reached is that, for various reasons, a considerable wealth of projects has been built up in Venice, even if it is enough to run through the list of cases to realise that in many of these, and particularly in those of a larger scale and which have more associations with environmental protection in a broad sense, the actual realisation of the project is still a long way off. It is not necessary to decide here whether this wealth in planning initiatives is a consequence of the financing from central government and from the European Union or, rather, whether it reflects the heightened awareness of the city's and lagoon's most serious problems, at least since the flood of 1966. In addition, naturally, this does not mean that other projects must not or cannot be created, for example, in terms of the lagoon environment's utilisation even for tourist purposes or the conversion of the production system. The fact remains, however, that the main problem which the system of government has to tackle is not so much the idea generation process, which, as has been seen, works, as, the decision-making and implementation process, which produces numerous difficulties.

Furthermore, the analysis of the projects shows how the network which governs Venice is fragmented, congested, at a very low density (in the sense that there are much fewer instances of constant collaboration between actors than would be potentially possible) and highly complex (in the sense that in each specific case the actors belong to many different territorial levels and to many different categories.) The hypotheses we have formulated to explain the significant involvement of the European Union and which stress the need to increase the political legitimacy of the projects and to introduce strict deadlines at the procedural level, reinforce this conclusion further.

If we now move on to analyse more closely the network of actors, it is seen that the only truly central actor in the government of Venice is the municipality, which goes far beyond its institutional responsibilities, even if it must be immediately noted how the municipality cannot really be considered a single actor. Particularly weak, on the other hand, is the representation of actors which can connect Venice to its territory (the Province and the Region, above all). Of greater importance, however, and not only for reasons linked to the financing of innovation, seems to be the presence of the state bureaucracy and of its peripheral offices. From this perspective, Venice would appear to be in a particularly fragile position in any potential change towards a federal system in Italy, something proposed from various sides.

The governance/planning model which has ruled the special legislation appears to have essentially failed. To support this conclusion, it is enough to consider that, in a situation in which most of the fragmentation is of an inter-institutional nature, it is important, but not surprising, to observe how the presence of collective bodies explicitly devoted to co-ordination (the *Comitatone* and the Protection Commission) does not actually manage to reduce the network's overcrowding but, in actual fact, simply adds a new decision-making channel to already existing ones.

Nevertheless, these considerations must by no means be interpreted in the sense that the system of governing Venice is excessively complicated and that it needs to be simplified, perhaps by means of unifying responsibilities or by the creation of new technical, political or administrative authorities. On the contrary, the subsequent analytical steps, which we had to make in order to find the connections, show how the real network which can be identified is extremely impoverished in the sense that interests and points of view which are important and certainly exist in the city are not represented within it. The weakness of the institutions which link Venice and the hinterland of the Veneto has already been mentioned above, as too has the absence or the scarce presence of institutions or bodies linking the city and its economic activities (the Chamber of Commerce, banks, business associations and, finally, the Port Authority itself). However, what is more important is that 'society' seems to be represented almost exclusively by environmental organisations. This is a factor which is worth highlighting because, as will be seen, this justifies, at the political level as well, the need or the opportunity at least to explicitly make the protection and valorisation of the environment a subject in any process of strategic planning. However, this monopoly of social representation also presents the problem of how to succeed in involving other subjects in the construction of the vision, especially the economic operators. Moreover, a further confirmation of the limitations in the government of Venice comes from the nature of the processes which are barely political, in the strict sense; i.e., that the parties (or trade-unions) never stand out as real prime players in the projects for transformation and, given the sources used, it would have been legitimate to expect it.

In conclusion, it can be stated that the largest obstacle at the political-institutional level and, more generally, in the government of Venice does not lie so much in the over-complexity of the process. On the contrary, it lies completely in the fact that the network of actors does not seem able to fully represent the complexity of society.

5. The construction of a shared vision: elements for institutional planning

It is now time to sum up the analyses made and to establish various elements of a planning procedure for solving the problem which was set in the introduction: the establishment of a possible process for constructing a Local Agenda 21 in Venice.

The first suggestion is that the choice of Local Agenda 21 as a decision-making procedure which brings the issue of Venice's development to the attention of the communities concerned seems justified, even with the considerable differences which can be found. Indeed, however much Agenda 21s might usually pursue other objectives and might show, in general, scarce consideration for the problem of development and, even more so, for the definition of a new development model, in light of the extreme complexity of the Venetian decision-making framework, of the considerable presence of the environmentalist component within it, but particularly of the exceptional importance of the natural and anthropic environment, the consideration of environmental issues actually seems to be an opportunity, rather than a constraint, which should be exploited in order to stimulate comprehensive appraisal of the lagoon city's future.

From this perspective, if it seems conceptually correct and useful in terms of communication to define Local Agenda 21 as the type of process for constructing a long-term shared vision which could succeed in drawing together environmental sustainability and the strength of the social and economic development model, it must also be remembered that it would be a procedure with a very different objective from the one pursued in Italy and abroad. Therefore, if objectives for Venice's global repositioning are accepted, to refer to Local Agenda 21 methods and techniques alone, risks obtaining a purely nominal process. Indeed, if a participative and communicative process is usually justified by the objective of changing the population's lifestyles and their consumption patterns, for example, in the Venetian context this must have other justifications and probably other features as well.

The second lesson which is obtained from the analysis of strategic planning experiences is that a process which tends to establish a long-term shared vision and to base concrete action programmes on this, cannot be based on an objective vision of the problems nor on a rational choice of solutions. On the one hand indeed, the use of knowledge, of any kind of knowledge however it is produced, within an interactive process of government is not the monopoly of a particular category of actors, i.e., the experts, and even less so of the institutional structures. On the other hand however, if the experts and the technical bureaucracies want to have their knowledge count in the process, and this is often a condition for the success of the whole exercise, they have to be able to present themselves as real key players and, from here, to draw up strategies based on reconstructing the network of actors, on anticipating their respective behaviour, etc. So, their choice is between playing the role of the professional (who musters his knowledge out of respect for deontological principles in favour of a specific interest) and of playing, in person, the role of an actor in the scientific community, indeed, in the various scientific communities.

The third consideration is that, even leaving the complexity of the subject aside, the Venetian situation seems difficult to manage, as the above-mentioned analysis of the networks of actors shows. To govern Venice is particularly difficult, not only and not so much for the most obvious reasons, so much as because the interaction between centre and periphery and, more generally the multilevel nature of policy making, multiply the risks of overlapping veto powers and of stalemate in decision-making. The multiplicity of criteria for action, of the type and dimension of interests in play, of the cultures and of the conceptual frameworks of reference, makes it extraordinarily difficult to solve the important problems in a satisfactory way, given also the network's low density and the weakness of the central actors.

From this perspective, to set a process in motion which aims to construct a shared vision means making an important and, in some ways, obligatory choice. This basically means attempting to make a different approach take priority in the government of Venice; one which is not based on the attempt to impose authoritatively choices deemed necessary within a given hierarchy of institutional actors who are linked to each other in various ways. On the contrary, the procedure suggested here consists in the adoption of a governance style which is centred on the search, agreed upon by a broader range of actors, for the definition of the problems and for the formulation of acceptable solutions, also and especially in terms of consensus which it is possible to reach on these. Such a model bases its very effectiveness on the indirect and truly strategic capacity to modify perceptions and behaviour of the actors which act within it. This appears to be a valuable opportunity for trying to make significant progress in a situation which very often runs the risk of stalemate in decision-making, despite the wealth of planning initiatives. Two further topics can be adopted in this respect.

The first is that Venice is a particularly important case, because it is essentially not an obvious case, of the failure of the intervention model based on inter-institutional co-ordination (the *Comitatone* and the Protection Commission) and on planning of a rationalist matrix, which departs from a political-administrative identification of the problems and tries, on this basis, to find the most suitable solutions. If the classic inter-institutional co-ordination allows for the problem of the difficult task of making the objectives which are pursued at the various territorial levels compatible and for the problem of the multiplicity of sectorial criteria, planning compensates for the inevitable long-term scientific unreliability and the limited nature of knowledge and of the models for its treatment. In this situation, it could well be in Venice itself that a radically new way of governing could be tried out and a process for defining a shared vision for sustainable development could be begun.

The second topic is that several attempts have already been made in this respect even in Italy and in situations where their success appears even less likely. We refer to the case of the so-called territorial pacts which in many southern areas attempt to create alternatives for credible local development, particularly after the end of the 'extraordinary intervention'. Certainly, these refer essentially to the theory of collaboration between social groups rather than to the theory of strategic planning, and they certainly set themselves much less ambitious objectives compared to the one of constructing a shared vision for sustainable development in Venice. Nevertheless, these experiences seem to show that the only possible course is that of planning processes which would encompass, and would not try to deny, the complexity of society.

It must also be said that a course of this kind involves costs which are by no means negligible. For example, this course is essentially an alternative to the one which is, alas, much used in our country, of resorting to a notion of emergency in order to justify special systems and shortcuts in decision-making. Evidently, the participants in the procedure for strategic planning must agree on the need to reverse the natural trends of development in Venice, also by means of a series of readily supported decisions, as well as on the urgency to act immediately in order to introduce the necessary correctives. Nonetheless, the efficacy of the procedure also depends on the ability to avoid impatience and excessive haste in wanting to cling to the first configurations of problems and solutions. As said above, the procedure's effectiveness lies in the fact that it changes the participants' perceptions and behaviour in an indirect way; that is, that it stimulates their learning. From this point of view, it therefore needs both a certain amount of time as well as actors equipped with enough perseverance and with the capacity to participate in a way which is intentionally ingenuous. Therefore, this procedure must not become some kind of super-government for Venice, but must develop in parallel with the everyday situation of government which must certainly continue in the practice of strategic planning for the whole of the necessary period as well, and also after its conclusion.

Having assumed that the opportunity to launch a process of strategic planning for Venice is agreed upon and that such a process is called Local Agenda 21 for Venice, since it aims to define a shared vision for long-term sustainable development, what are the conditions for such an initiative's success?

An initial condition for success is related to the definition of the problem, particularly in terms of the nature of the decision-making network involved. Indeed, on the one hand it is clear that the multilevel nature of the process cannot realistically be denied since the non-Venetian actors already constitute an essential part of the decision-making network. On the other hand, however, focusing on Venice's speciality (and not only of its historical centre but also of its specific mix of economic activities – heavy industry, tourism and the port – and of its natural environment – the lagoon) risks defining the problem in a way which is far too restricted. This runs the risk of hindering a co-operation/collaboration process between the actors because of the fear that, in the overlapping of criteria for action, some interests might become unilaterally sacrificed.

In some ways, the Venetian situation is a classic one, in the sense that it is a pure case of the prisoner's dilemma in which co-operation is made difficult not only by the lack of communication channels but, particularly, by the lack of comprehension that there exists a single position which can become a non-zero-sum-game. In a deliberately exaggerated way, it could be said that, in order to set the process in motion, it is probably necessary to deny Venice's special nature because such a definition of the problem, far from establishing common ground, evokes political relationships (in the broad sense) which are already too rigid. In this case as well, the example can be made of the Southern question where the definition of its special status, which was the basis of extraordinary intervention, has triggered a process of policy making which is substantially self-referential and extremely poor.

Therefore, it is necessary to find a definition for the problem which can attest to the discontinuity with efforts which were made in the past and towards which, today, a sense of disappointment tends to prevail.

The first proposal for avoiding the stalemate, which we have already mentioned, is implicit in the initial choice to consider environment and development simultaneously. Indeed, to introduce environmental considerations in the broadest sense possible must be considered an opportunity rather than a constraint because, if it is true that the key to solving many problems lies in their separation, on the contrary, the construction of a shared vision is, paradoxically, facilitated by the existence of functions which are in conflict with one another.

However, this first proposal still does not seem to suffice and, so, the second, and more radical, proposal which can be advanced is one to define the problem of Venice's sustainable development within the scope of global threats which the current development model has to take into account. We refer not only to climate change, which also challenges Venice's problem both directly and at its root, but also to the current trends in several of Venice's key sectors (the chemical industry, tourism, port operations, etc.), or rather to the challenges implied in the social and economic dynamics of Eastern Europe. Therefore, a Local Agenda 21 process should begin by asking what the consequences of the changes are for Venice, consequences which the city cannot influence anyway, and, from here, begin to try and identify a development procedure, thereby recovering the city's speciality at this level. In order to reduce the transaction costs between the necessary participants (which, as we shall see, are numerous) it seems appropriate to try to reposition the issue by looking outside Venice rather than at its historical centre and/or the lagoon. If it were possible to put such a conceptual operation into practice, the base would be established, if for nothing else, for a procedure which differs from the ones the city has already experienced and which are regarded by now with considerable disillusion. Furthermore, such a conceptualisation of the problem would recover a dimension, i.e. of Venice as a global city, which is deeply etched in the city's history and, in some ways, in its genetic code.

However, these considerations immediately lead to tackling the second factor for the proposal's success. Indeed, a similar redefinition of the problem is closely linked to the fact that the process initiator should be really credible in confronting both this problem and, above all, the global/local connection. The starting phase of the process, in fact, is particularly delicate because it is necessary to succeed in overcoming the inertia and the mistrust of the required participants towards any prospective change in the status quo. Therefore, the initiator's credibility means, essentially, his reassurance that the objective pursued is not one of altering the present distribution of roles and resources in favour of one subject and/or to the detriment of another. From this point of view, the international experiences and the situation of the governmental network in Venice cause more doubts than they resolve. Abstractly, it would be possible to imagine either that the initiative could take place at the local level (as happens in the case of Local Agenda 21s in which the key role is often played by the municipality and the demand for initiation originates with the environmentalist world), or that the initiative could take place at the national level (as, for example, in the case of strategic planning for environmental issues in Holland), or, finally, that it takes place at the intermediate or regional level (as has often happened in cases of coastal zone management).

Each of these three solutions has both negative side-effects and potential advantages:

(1) At the local level, there is the advantage of relying on the only actor which, as has been seen, is at the centre of the innovation processes and in the transformation projects. However, there is a risk of losing the problem's definition, that is, the global dimension, which seems necessary for guaranteeing a different process from the past;
(2) At the national level, there is also a tradition to be used as a base, but this involves the danger of going through the attempts of hierarchical government again (Venice as a national problem) which appear largely inadequate for confronting the problem;
(3) At a regional level, on the other hand, there is no tradition of intervention and development which is a deficiency which is very difficult to make up for. However, there would be the advantage of including the Venice question in the moves towards transforming the State's structure into a federalist one.

In short, there seems to be only one solution which can be employed in the short-term. However, it is by no means completely satisfactory because it risks 'diplomatising' and over-institutionalising a process which is, already in itself, very fragile. It is the solution of having a collective body play the role of process initiator; a body which could encompass all the levels of government; i.e., the *Comitatone*. However, the conditions are that it would begin:

(1) with an explicit and open recognition of the limits and indeed of the real failure in relation to the problem which is to be dealt with, of the traditional model of intervention and, at the same time
(2) with the proposal for an approach which is radically different from the issue: a pact for Venice, which could place all the participants on an equal footing, irrespective of the dimension of the interests represented or of their nature, whether political/institutional or social.

The collective nature of the initiator of the process of Local Agenda 21 introduces immediately what appears to be the third condition for success. The development and the efficacy of a Local Agenda 21 with the characteristics shown above depend strictly on its capacity for really involving all the actors necessary for governing Venice. From this perspective, the present decision-making networks should be substantially enhanced since they reflect the required complexity only very partially, even if in the past this was not, perhaps, the case. In fact, the government of Venice today seems to be made up almost exclusively of political–administrative institutions with the single addition of environmentalist interest groups. Therefore, there is a deficiency in representation both in terms of culture and experts and in terms of civil society, as well as and above all, where economic interests are concerned.

This last point is particularly significant since not only do the official representatives of the business world appear singularly absent from the debate, but so do Marghera's heavy industry and the representatives of the port, and they rarely make their voices heard. Not to mention as well the tourist industry which does not even seem to be able to equip itself with a representative organisation which can have its own say on projects and problems which concern the industry's economic interests very closely.

This is a key point on which it is important to be clear. The possibility of starting a process of strategic planning and for constructing a shared vision of sustainable development cannot exist unless there are powerful actors inside such a process who see the advantages in reducing the future's uncertainty for their own interests and for the definition of their own strategies. The third and, perhaps, most difficult of the conditions for the success of such a process consists in the capacity to substantially enhance the network through a full involvement of the economic interests. In the absence of such involvement, particularly of the tourist industry which forms such an important part of the Venetian economy and life, the whole undertaking risks being essentially irrelevant or, in any case, not taking off because it is not seen to move away from similar efforts made in the past and which, whether rightly or wrongly, are branded as idealism.

The fourth and final factor for success concerns the process's design in the strict sense: that is, even if the network were enhanced and representative, if a credible initiator were found, and the problem were redefined in global terms, the undertaking could still fail due to the lack of adequate operational procedures. However, the international experience which we have mentioned, albeit briefly, is already sufficient for furnishing the key elements, and, most importantly, the necessity to leave the definition of the rules of the game to the first phase of interaction between the actors. The other elements, such as for example, the opportunity for having facilitators, the controlled opening nature of the process, the need to provide monitoring and evaluation instruments which are agreed upon and reliable, etc., are, in some ways, already part of a state of the art which need only be mentioned. Among other things, it is also necessary to emphasise how, despite everything, the cognitive deficiency is lower in Venice than it is elsewhere, given that a vast series of studies and research projects have been recently carried out which have documented all the various aspects of the complex situation of the lagoon city. Rather than carrying out further research or investigations, it seems necessary to socialise and to communicate the basic facts which, although they have significant uncertainties as regards the interpretative models, seem available nonetheless. From this viewpoint, as we have just said, an essential chapter in the planning of decision-making procedures concerns the endeavours to communicate which should be made in order to involve the population in the procedure. This is an aspect which is rightly considered essential and which would also have the advantage of going beyond the current situation which sees the representation of society monopolised by environmental groups alone.

Summarising, therefore, the main planning elements which can be identified are the following:

(1) It is necessary to define the procedure of Local Agenda 21 as an exercise of long-term strategic planning and, in particular, of constructing a shared vision for the development of Venice.
(2) It is appropriate to consider the environmental dimension as an integral part of the problem and not only as a constraint to possible solutions.
(3) It is necessary to deal with Venice within the global dimension and, that is, to consider the problem of Venice's sustainable development within the sphere of global threats of a social, economic and environmental nature, which the present development model must take into account.
(4) The role of process initiator must be taken by all of the institutional actors involved, probably within the framework of the existing inter-ministerial committee, as a definitive break with the past and as the proposal of a new pact for Venice.
(5) Every effort must be made to enhance the network of actors involved and, in particular, to give a voice to and to make room in the process of strategic planning for the economic interests and, more generally, for all the representatives of Venetian society.
(6) It is necessary to design the operational procedures carefully, basing them on the best examination possible and on the most advanced experiences abroad.

These are elements which certainly define a difficult challenge for planning also, and above all, because of the considerable level of discontinuity in relation to the existing situation which these would involve. However, we believe that no-one would be able to imagine that governing Venice's future could be deemed a simple exercise.

BOX 1. *ROM policies in Holland*

Within the Dutch planning system, an extremely innovative policy has been proposed for eleven areas where there was an evident need for reconciling the demands relating to local economic development with requirements linked to the protection and valorisation of the environment. These areas are known in Holland as ROM areas (*Ruimtelijke Ordening en Milieu*, territorial and environmental planning). The approach, begun between 1986 and 1988, has produced a number of action plans which have already passed through the formal procedure and others which, instead, are still in the course of development.

The approach proposed for these areas, the selection criteria and the design of the whole process constitute a novelty in Dutch planning which, traditionally, was divided in a hierarchical and sectorial way, and elaborated within the public administration, with few openings for other actors. This change in approach departs from the recognition of the fact that the traditional planning system is ineffective for environmental matters, not only in terms of the search for solutions but also in terms of the very definition of the problem.

In this case, the public administration acted as the promoter for an initial phase of the process, and then set up opportunities for negotiation with all the interest groups relevant for that area: not only the different levels of government but also businesses, trade associations, non-governmental organisations, etc.

The eleven areas whose development was deemed a matter of national importance can be divided initially into two categories. On one hand, there are some areas within which the whole environmental system is seriously compromised and the plan is seen as a necessary step in order to avoid prejudicing future development. On the other, areas are found which are characterised by a very high environmental quality. Here, the main aim is to look for effective instruments for non-restrictive protection in order to keep the natural heritage and the total ecological value intact. Other distinctions can then be made on the basis of how important the areas are in relation to networks of local relations, both national and supra-national, given that the two areas of Schiphol airport and the port of Rotterdam have an importance which is incomparable to that of some rural or natural areas. As a result, the choice of areas was made by focusing on different criteria: the national importance, the existence of different environmental problems, the wish to include both 'clean' and 'dirty' areas, priority for areas where an administrative impasse had been encountered. The final choice was made in consultation with the provinces concerned.

The stages which the processes were divided into, even though they often differed from one case to another, can be put into a general scheme which shows a starting phase, a phase of developing the action plan and an implementation phase.

In the starting phase, one or more partners takes the lead in the co-operation process, during which the stake-holders are invited to participate. Once it is clear who the participants at the negotiating table are, they sign an initial agreement which establishes the rules of the game, specifies the general objectives, the content and the organisation of the process of developing the action plan and defines the timescales and resources required.

In the phase of developing the plan, several different stages can be distinguished: an initial stage devoted to the contruction of a common scenario for future development (or of more scenarios, even if this risks leading to a block in the process), followed by a stage of examining the main issues during which the participants are often requested to swap roles; finally, there is a stage of constructing action proposals which can indicate ways of territorial and economic development.

Ultimately, after final agreement between the participants on the action plan (which, sometimes, is not signed by all the participants) and after the passage through the formal procedure, the implementation phase begins, in which each of the participants in the development phase takes on specific responsibilities.

In all the ROM areas the most effective structure has been of a bottom-up kind, except in the case of the central area (the Randstad green heart) in which the approach was, rather, top-down. Here, the approach actually began with the strong desire at the state level to preserve this green area which is located at the centre of various metropolitan areas in expansion. In an initial phase, therefore, the municipalities were kept outside the project and were only called upon to participate subsequently. In this way, the time for the municipalities to become involved was when the plan already provided for something more than the simple obstruction of all residential expansion in the area which was the main point of disagreement with the municipalities concerned. In this way, it was possible to reach an all-win solution for all the participants.

An important innovation consists in the emphasis placed on the implementation of the plans. The importance of the implementation stage is linked to the need to maintain involvement and dialogue between all the actors. Without this, the commitment which is necessary for progressing from the development phase to the operational phase is lost.

About the Authors

IGNAZIO MUSU, full professor of Political Economics at the University Ca' Foscari of Venice and president of the Scientific Committee for the Fondazione Eni Enrico Mattei in Venice.

MAURO BON, custodian at the Venice Civic Natural History Museum for vertebrate zoology. He specialises in terrestrial vertebrate ecology and fauna management.

DANILO MAINARDI, full professor of Nature Conservation in the Environmental Sciences degree course at the University Ca' Foscari of Venice. His research concerns the ecology and the behaviour of aquatic and terrestrial vertebrates.

LUCA MIZZAN, custodian at the Venice Civic Natural History Museum for marine biology. He specialises in marine and lagoon benthos ecology.

PATRIZIA TORRICELLI, associated professor at the University Ca' Foscari of Venice, holds the professorship for the Basics for the analysis of ecological systems in the Environmental Sciences degree course. She carries out research on the ecology and behaviour of lagoon fish species.

ANDREA RINALDO, full professor of Hydraulic Constructions at the University of Padua.

CARLO MAGNANI, architect, associated professor of Architectural Composition at the Venice University Institute of Architecture.

TRAUDY PELZEL, architect, carries out research at the Venice University Institute of Architecture.

MAURIZIO RISPOLI, full professor of Business Strategy at the University Ca' Foscari of Venice and Rector of the same University.

FRANCESCO DI CESARE, research collaborator at the Department of Economics and Business Management at the University Ca' Foscari of Venice.

ANDREA STOCCHETTI, research doctor, contract professor of International Marketing at the University of Ca' Foscari and research collaborator at the Department of Economics and Business Management.

JAN VAN DER BORG, teaches the Economics of Tourism at the Universities of Venice and Rotterdam.

ANTONIO PAOLO RUSSO, PhD student at the Tinbergen Institute at the Erasmus University Rotterdam and researcher at the European Centre for Comparative Urban Studies of Rotterdam.

ENZO RULLANI, full professor of Business Strategy at the University Ca' Foscari of Venice.

STEFANO MICELLI, researcher in Industrial and Commercial Practice and lecturer in Marketing at the University of Udine.

BRUNO DENTE, teaches Public Policy Analysis at Venice University Institute of Architecture and carries out research at the Institute for Social Research in Milan where he is also one of the scientific directors.

CINZIA GRIGGIO, was a researcher at the Fondazione Eni Enrico Mattei (Venice office) until 1997.

ANDREA MARIOTTO, research doctor of Territorial Public Policy, he works as a consultant in matters concerning planning and the analysis of public policy.

CAROLINA PACCHI, carries out research at Milan polytechnic in the field of territorial planning.

Index

FONDAZIONE ENI ENRICO MATTEI (FEEM) SERIES ON ECONOMICS, ENERGY AND ENVIRONMENT

This series serves as an outlet for the main results of FEEM's research programmes in the areas of economics, energy and environment.

The titles published in this series are listed at the end of this volume.

FONDAZIONE ENI ENRICO MATTEI (FEEM) SERIES ON ECONOMICS, ENERGY AND ENVIRONMENT

This series serves as an outlet for the main results of FEEM's research programmes in the areas of economics, energy and environment.

The Scientific Advisory Board of the Series is composed as follows:

Kenneth J. Arrow
Department of Economics
Stanford University, Stanford, California, USA

William J. Baumol
C.V. Starr Center for Applied Economics
New York University, New York City, USA

Partha Dasgupta
Cambridge University, Cambridge, United Kingdom

Karl-Göran Mäler
The Beijer Institute, Stockholm, Sweden

Ignazio Musu
University of Venice, Venice, Italy

Henry Tulkens
Center for Operations Research and Econometrics (CORE)
Université Catholique de Louvain, Louvain-la-Neuve, Belgium

Domenico Siniscalco (Series Editor)
Director, Fondazione Eni Enrico Mattei, Milan, Italy and
University of Turin, Turin, Italy

Carlo Carraro (Series Associate Editor)
Fondazione Eni Enrico Mattei
and
University of Venice, Venice, Italy

The titles published in this series are listed at the end of this volume.